PROFESSIONAL VERIFICATION
A Guide to Advanced Functional Verification

PROFESSIONAL VERIFICATION
A Guide to Advanced Functional Verification

PAUL WILCOX
Cadence Design Systems, Inc.

Kluwer Academic Publishers
Boston/Dordrecht/London

Distributors for North, Central and South America:
Kluwer Academic Publishers
101 Philip Drive
Assinippi Park
Norwell, Massachusetts 02061 USA
Telephone (781) 871-6600
Fax (781) 871-6528
E-Mail <kluwer@wkap.com>

Distributors for all other countries:
Kluwer Academic Publishers Group
Post Office Box 322
3300 AH Dordrecht, THE NETHERLANDS
Telephone 31 78 6576 000
Fax 31 78 6576 474
E-Mail <orderdept@wkap.nl>

 Electronic Services <http://www.wkap.nl>

Library of Congress Cataloging-in-Publication

Title: Professional Verification
       A Guide to Advanced Functional Verification
Author (s): Paul Wilcox, Cadence Design Systems, Inc.
ISBN: 1-4020-7875-7
ISBN: 1-4020-7876-5 (eBook)

This book is dedicated to my wife, Elsa, for her unwavering love and support, and to my children, Jonathan and Elizabeth, for the inspiration and joy they have brought to my life.

Contents

Authors

Paul Wilcox is the Director of Methodology Engineering at Cadence Design Systems, where he has worked since 2002. Previously, he worked at Cisco Systems, StratumOne Communications, 0-In Design Automation, and Sun Microsystems. He holds patents for work in advanced ASIC design and verification. Paul holds a Bachelor of Science degree in electrical engineering from Northeastern University and an MBA from San Jose State University.

Kurt Johnson is group director of Custom IC Marketing for Cadence Design Systems. At Cadence, Johnson has served with AMS Design Environment Services, where he established the AMS design environment from the ground up. He was also principal technical architect and strategist for full custom methodology services at Cadence IC Methodology Services. He has worked for Qualcomm, Western Digital, Teledyne Systems, and Motorola. Johnson earned his BSEE from Purdue University.

Ray Turner is the senior product line manager for Cadence's Incisive Palladium accelerator and in-circuit emulation systems, part of the Incisive Verification Platform. Before joining Cadence, he was the EDA marketing manager for P CAD products for seven years. Overall, Ray has 18 years experience in product management for EDA products. He also has 14 years experience in hardware, software, and IC design in the telecommunications, aerospace, ATE, and microprocessor industries. Ray received his Bachelor of Science degree in electrical engineering from Loyola University of Los Angeles. He holds patents for early work in digital signal processing and has authored two books on engineering.

Acknowledgements

Sir Isaac Newton once remarked, "If I have seen further [than certain other men], it is by standing upon the shoulders of giants." This book is based on the experiences and hard work of many giants in the design and verification of modern ICs. It would be impossible to list all the individuals who have contributed to the collected knowledge contained in this book, but it would be foolish to not acknowledge their contribution.

I have encountered many "giants" in my career who have taken the time and had the patience to teach me much of what is contained in this text. For that I would like to acknowledge the friends and co-workers I have worked with at Sun Microsystems, 0-In Design Automation, StratumOne Communications and Cisco Systems. Special thanks to Willis Hendly, David Kaffine, James Antonellis, Curtis Widdoes, and Richard Ho.

This book is the product of the efforts of many people at Cadence Design Systems, and I would like to acknowledge the following for their contributions and efforts in reviewing the text: Andreas Meyer, Grant Martin, Leonard Drucker, Neyaz Khan, Phu Huynh, Lisa Piper, and the entire Methodology Engineering team.

I want to acknowledge Linda Fogel for her tireless and professional editing, along with Kristen Willett, Kristin Lietzke, and Gloria Kreitman in Cadence's marketing communications group.

A special acknowledgement to Paul Estrada for providing me the opportunity and time to write this book and for showing faith in me when even I was ready to give up. One could not ask for a better mentor or friend.

Finally, I would like to acknowledge the true giants of my life, my parents, Eleanor and Gary Wilcox, for their love and support, and for teaching me the nobility of education.

SECTION 1
THE PROFESSION OF VERIFICATION

Chapter 1

Introduction
Thinking about how it might not work

After years of doing what I considered grunt work in test, tool development, and verification, I finally got my chance to design a major portion of an important chip. I had created a detailed specification and beat all the scheduled milestones. My design was meeting its performance goals with time to spare, and the initial layout looked great. And then, two weeks before tapeout of the entire chip, the bug reports began to come in. The random verification regressions had been running fine for weeks until some of the parameters were loosened. Suddenly, my block was losing or misordering transactions, and all the simulations were failing. I found what I thought was a one-in-a-million corner case bug, but the next day the simulations were failing again. Another fix and another fix and still the bugs kept popping up. I was called in by the project managers. The tapeout deadline was at risk of slipping and it was because of me.

As I drove home that night, I tried to figure out what was going wrong. I had followed all the design rules, creating a very complex design in smaller size and greater performance than had been required. The data structure I had come up with could support many advanced features, and we were even patenting it. As I thought about it, I saw that the complexity I had added also created many new possible side effects, and the simple testbench I had written could not test these side effects. I realized that the only way I could get this design back on track was to stop thinking about how it should work and start thinking about how it might not work. It was at this moment that I began to understand functional verification.

But I'm not alone in going through this. Most engineers throughout the integrated circuit (IC) industry have had similar experiences. Fortunately, functional verification is evolving from an afterthought to an integral part of the development process. The evolution has occurred not because of forethought and careful planning, but out of necessity. Functional verification teams must keep up with growing complexities, growing device sizes, rapidly changing standards, increased performance demands, and the rapid integration of separate functions into single systems. Functional verification of today's nanometer-scale, complex ICs requires professional verification.

This book explores professional verification in a practical manner by detailing the best practices used by advanced functional verification teams throughout the industry. The goal of this book is not to present research into new areas of verification or to provide a how-to manual for a specific tool or language. Instead, this book describes the advanced verification process based on what real teams are doing today so that you can incorporate this information into your own work.

LEARNING FROM OTHER'S BEST PRACTICES

Anyone who is even indirectly associated with the development of advanced ICs today is well aware of the numerous issues surrounding functional verification. Larger designs with increasing complexity and shorter development cycles have made yesterday's basic verification techniques unusable. While verification teams struggle with doing more verification to ensure higher quality and first past success, they are faced with fewer resources and outdated tools. Functional verification is a moving target for many teams who feel they are going to battle unprepared and understaffed.

Where can verification teams go to address the issues they face day in and day out? Most new research in verification focuses on mathematical techniques or system-level approaches that require major changes in verification as well as design and infrastructure. Engineering schools generally do not include functional verification in their curriculums. The only place to turn to for guidance is to other verification teams. By studying the best practices of advanced verification teams and learning from their successes and failures, engineers can gain valuable insights for addressing their own verification needs.

There is no perfect verification team in the industry today, and no two teams face the same set of issues. Therefore, it is important to examine the best practices of a wide variety of teams to glean the issues most important to you. Cadence Design Systems works with and studies the practices of the most advanced verification teams. This book is based on these collected experiences. It provides a wide array of knowledge and practices that you can use to address your specific concerns.

IS THIS BOOK FOR YOU?

If you are a researcher or an engineering student, this book offers a useful presentation of the advanced techniques used by today's successful verification teams. However, the book assumes some basic knowledge in the known issues and processes used in industry today. A list of excellent resources on the fundamentals and techniques of functional verification is provided in Appendix A, "Resources."

If you are a project manager who oversees IC development, you will find the information on methodology and process improvements valuable. If you are an architect, system designer, system integrator, software developer, or tool developer, you might not consider verification a primary concern. But gaining a good understanding of the overall verification process and how that process can be integrated into your own area of expertise could be beneficial. Often design engineers verify their own block or another designer's block. Many of today's register-transfer level (RTL) designers understand the importance of functional verification and the role it plays in the development process. This book provides information on advanced verification that you can use in your job as you move through your career.

But the main audience for this book is the verification engineer. You are the professionals who face the day-to-day responsibilities and challenges of verifying the largest and most advanced ICs today. This book is an invaluable resource for understanding the complete verification process and how advanced teams throughout the industry successfully attain their goals.

This book focuses on the practical application of advanced verification techniques using today's best practices. The first section presents some of the fundamental topics in verification today and discusses the profession of verification. The second section shows how verification issues can be addressed, and shows how all of the practices fit together to form a complete unified methodology. The final section provides an in-depth look at many of the topics discussed in the second section.

Chapter 2

Verification Challenges
Missed bugs, lack of time, and limited resources

Every development team faces verification issues of some kind. Some might be due to the size or number of designs, some to the complexity of the design, some to the verification process being used. Verification teams continually attempt to address these issues only to find that new problems arise that are more complex than the original ones. Almost every issue in verification today can be placed in one of three buckets: missed bugs, lack of time, or lack of resources. The most pressing issue is the inability to find all the bugs during the verification process. Given enough time and resources, teams can verify with a high degree of confidence that a design will work. But having enough time to complete verification is a challenge. So more resources, more compute power, and more processes are thrown at the problem in hopes of decreasing the time to complete successful verification. Yet resources are expensive, which leads to the perception that verification takes too many resources to be successful on time. Teams need to find all the bugs in the shortest amount of time, with the fewest number of resources, in the most efficient and effective manner. Let's take a closer look at each of these areas.

MISSED BUGS—ARE WE JUST UNLUCKY?

The highest priority for verification teams has always been to find bugs. All the effort placed in architecture, design, implementation, and verification can be wasted with one missed functional bug. The farther down the supply chain a bug is found, the more costly it is to everyone involved. You can determine a company's belief in their design and verification processes by examining the number of planned spins they account for in their schedules and budgets. Some teams automatically plan and budget with the assumption that a certain number of functional bugs will not be found in verification. Other teams plan for first-pass success, and respins are only part of the contingency plan and budget.

It does not matter how fast or advanced your process is if bugs slip through into silicon, causing the project to be delayed or cancelled. A recent study by Collette International Research of IC/ASIC designs and functional verification issues related to chip design found that the most common sources of functional bugs found in silicon are design errors, incorrect or incomplete specifications, and changing specifications.[1] Each of these problems can be addressed with better use of tools and verification processes. Designer errors can range from simple typographical errors to incorrect implementations of complex functions or features. Using tools targeted at improving code quality and testbenches that thoroughly stimulate and check the design can increase the chance of finding these bugs. Incomplete or incorrect specifications are due to a lack of process and discipline in the development process. Verification can address this problem by requiring an executable specification and implementing detailed review processes. Miscommunication errors most often occur when parts of a design developed by different engineers or teams are integrated together. Verification can remedy this with thorough integration and system testing.

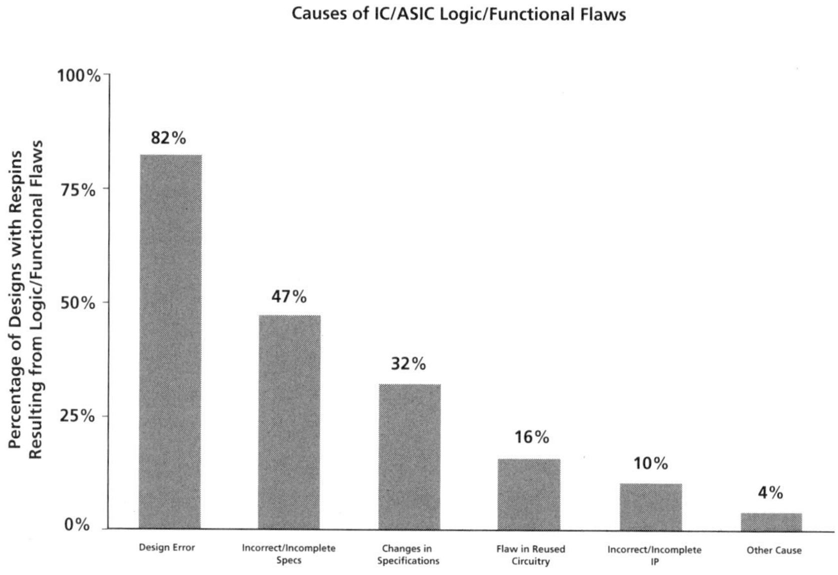

Causes of IC/ASIC Logic/Functional Flaws

Figure 1. Reasons for Missed Bugs from the Collette Study

Every project seems to encounter one or two "one in a million" bugs. No matter how thorough the verification process is, it is impossible to know about and verify every possible scenario. Inevitably, a bug is found in the lab that seems so obscure, requiring an unthinkable number of events to occur in just such a sequence. Engineers console themselves by thinking that there is no way they could have thought of that bug ever occurring. But when two or

[1] Collett International Research, Inc., 2002 IC/ASIC Functional Verification Study, North America.

three of these "one in a million" bugs are found in every project, a solution needs to be found.

The difficulty in finding obscure bugs in large designs is that there are more possible scenarios than there is time to test. Verification teams attempt to address these bugs using advanced automated techniques, such as formal verification or constrained random testbenches. Automating the processes increases the likelihood that the correct sequence of events to stimulate and catch the bug occur. But like any bug, finding obscure bugs is best handled by using new or better tools and processes.

THE NEED FOR SPEED

Time-to-market pressures have forced the entire IC development process to be completed in less time, while the size and complexity of designs continue to increase. Many verification teams are challenged by completing the verification process on time.

Probably the most asked question by development teams is "when are we done?" Verifying large complex designs is an exercise in risk management. If you stop and tape out too soon, you risk finding a bug in silicon. If you wait and tape out too late, you may have wasted an opportunity to get to market sooner. There is no easy answer to determining when you have done enough verification, and often the answer is different for each design. Teams have attempted to use metrics, such as bug rate or coverage, but they do not always provide accurate enough information. Teams often determine when they are done based on a mix of experience, metrics, and gut feel.

Teams may not be able to determine when they are done, but they usually know when they are not. Every team has a feel for what testing must be done before a design is ready to tape out. It is not until they reach the point that they have tested everything that they can think of that determining completeness becomes an issue. Often teams never reach this point due to schedule constraints—no need to address determining when you are done if you know there is testing still to complete. This issue often boils down to a need for verification speed. If teams are able to verify what they know needs to be tested as fast as possible, they will then have time to analyze metrics and make a thorough conclusion on completeness before the clock runs out.

Finding easy or obvious bugs late in the verification process is very frustrating. Having a functionally correct design is a prerequisite for many development processes, such as synthesis, software development, and system verification. Finding a bug late often means resetting these processes, as shown Figure 2. Obviously, the earlier bugs are found the better, but finding bugs sooner is only a benefit if it decreases the total verification time. Some techniques take so long to debug or to reverify that the total verification time is the same or worse.

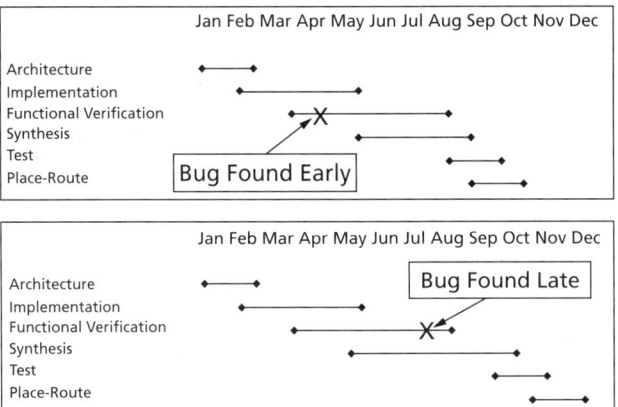

Figure 2. Late State Bugs Delay Project

Finding 98 percent of the bugs in a design is just a matter of using a methodical process. Finding the last 2 percent, however, often takes the most time and effort. Finding tough bugs is considered partly an art, partly luck, and a lot of hard work. Having engineers experienced in debugging is always helpful, but it still takes a lot of hard work. Tough bugs are hard to find because of misconceptions, because they involve a large amount of the design, and because they require many simulation cycles to find and recreate. Designers or debuggers often have misconceptions about the operation of the design, so they might not be focusing on the real cause. When debugging complex designs, the bug seems to move throughout the design—it starts in one block, is traced to another and another, until it is found in a place nobody would have thought to look. Once an incorrect operation is identified, it might take many more cycles to recreate the bug using the correct debug tools.

DOING MORE WITH LESS

Given enough time and resources most advanced verification teams can reach their goals. But time and resources are valuable commodities. Verification teams need to reach their goals in less time with the least amount of resources possible.

The verification process requires many resources, including engineers. Finding experienced verification engineers is difficult. Currently, schools do not offer formal training in verification, so teams are left with trying to hire experienced engineers or training inexperienced engineers. This often leads to specialized verification engineers who know how to perform some verification tasks, but not all. Specialized resources can only be utilized on specific tasks, so unless the process is managed very carefully, resources are not utilized to their fullest extent. The net effect is inefficiencies in the verification process, limiting the amount of work that can be done.

Verification resources also include compute power and verification tools. In the past, many teams have attempted to throw compute power at verification to get it done faster. Today, computer and networking hardware is relatively cheap, but outfitting a large server farm with the necessary software licenses and verification tools can be costly. Quite often verification tools focus on one specialized task in the process and can only be used for a small percentage of the overall project time. Outfitting a server farm with software to meet the demands for this limited amount of time can be very costly.

A mantra of many good code developers is "write once and use often." Unfortunately, the mantra in verification seems to be "write often and use once." Time is often wasted capturing and replicating the same information multiple times for different processes and tools. Different environments require the same information but in different representations. The goal of verification teams should be to reuse the information from task to task and from project to project. Verification reuse can provide a huge productivity gain if done correctly. Making models and components reusable requires more time and effort, but if this is amortized across multiple projects, it is often worth the effort and costs. Verification teams need to identify when reuse is applicable and have the processes and methodology in place for utilizing components or models multiple times. If teams do not plan for reuse, the work can be wasted.

Reusing design blocks is prevalent today in large complex designs. It is often easier to reuse or modify a block from a previous design than to design a block from scratch. Unfortunately, the verification for that block may have been done by a different team using different methods and environments than your team plans to use. Do you trust that the design was verified correctly the first time? Does the design need to work differently in your system? These questions often lead development teams to reverify existing blocks to reduce the risk of failure.

Development teams want to be able to reuse old design blocks without having to reverify the old design. This requires design teams to design blocks that work for the intended design as well as newer designs. Verification teams need to develop environments that can quickly reverify existing designs when they change and that can also expand to verify designs within different system environments. The gain in efficiency design teams realize from reusing design IP needs to be met with an equal focus on efficiency in the verification process.

FRAGMENTED DEVELOPMENT, FRAGMENTED VERIFICATION

Whether the verification issue is related to missed bugs, speed, or efficiency, addressing the problem involves careful analysis and proper execution of the solution. Problems involving missed bugs are usually related to not using the correct tool or technique or using them incorrectly. These types of problems are often resolved by understanding the root cause of the problem and then selecting the correct technique to address it. The third section of this book discusses different verification techniques and how they can be used to address these types of issues.

Addressing issues of speed and efficiency, however, is not as straightforward as addressing missed bugs. Fragmentation in the verification process is the major cause of lost speed and efficiency. Today's verification process is fragmented into many isolated stages that do not share information or common techniques. This fragmentation results in duplicated processes, incompatible techniques, and lost time.

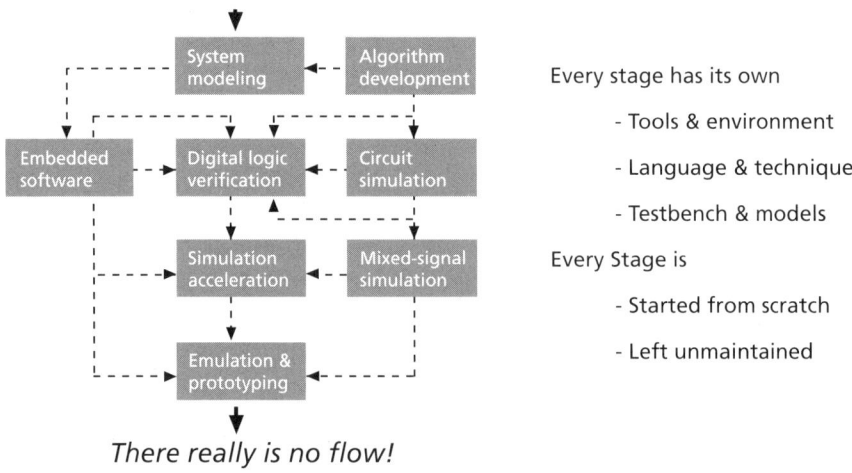

There really is no flow!

Figure 3. Typical Verification Flow

If we look at a typical verification flow, the fragmentation is obvious. Each task has its own stage with its own tools, environment, user interface, and models. Reuse from task to task, often known as vertical reuse, is limited or impossible. The same information is recreated at each stage, only to be left unmaintained once the task is completed. This "information rot" makes it nearly impossible to quickly make late changes in the design and rerun the task.

Fragmentation also exists from project to project. Few companies have a common verification flow for all projects. Even derivative projects often require new verification flows to be developed. Because each project is different, reusing models or information is impossible. Even though design IP can be reused from project to project, verification IP used to reverify the design often cannot be reused. Fragmentation also exists between design chain partners. Designs today are linked in a chain with IP developers providing blocks to IC developers, who provide devices to system developers. Fragmentation between design chain partners results in recreating verification environments at each stage in the design chain.

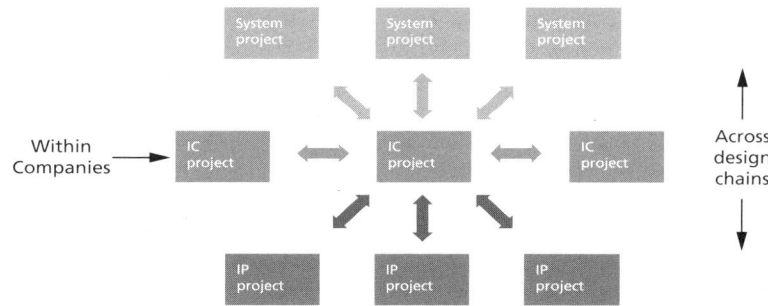

Verification environments & verification IP not reused

Figure 4. Fragmentation Across the Design Chain

The verification process has also become fragmented due to the ad hoc approach most teams use to develop their process. Instead of addressing the verification process as a whole, teams address individual issues on an as-needed basis. Teams add new techniques or tools without regard for the overall process, resulting in a flow that resembles islands of automation. With the advent of hardware description languages (HDL) and common implementation flows, the productivity of design teams has surpassed the ability for verification teams to keep up. Verification teams, who are under great time and resource pressure, end up just fighting fires and not addressing long-term issues, which results in fragmented approaches to verification.

No single tool or method can address the fragmentation in your development process. In fact, they might make it worse. What is needed is a methodology to unify all the stages, from system design to system design-in, across different design domains and projects. Only by unifying the entire verification process will fragmentation be removed, making dramatic gains in speed and efficiency possible.

The second section of this book looks at a unified verification methodology that is based on best practices used by advanced verification teams today. It gives you a view into how the different tools and techniques can be used together to address today's major verification challenges. But first, let's look at the role advanced functional verification plays in developing a unified verification methodology.

Chapter 3

Advanced Funtional Verification
Viewing verification differently

Advanced functional verification is not simply doing more of what you are already doing or using more powerful tools to make your job easier. Advanced functional verification is a fundamentally different way of thinking about and performing verification of large complex designs. Let's look at some of the basic principles of advanced functional verification and the implications these principles have in verification today.

VERIFICATION AS A SEPARATE TASK

For many years, verification was considered part of the design process. The verification strategy was developed by the design team, and verification was performed after the code was completed, usually by the designer who wrote it. This approach worked for small designs, where verification made up only a small percentage of the total project time. As the size of designs grew and the complexity increased, verification became a much larger portion of the total development effort. Advanced development teams realized that they had to treat verification as an independent development task.

Separating the verification of a design from the development of the design helps improve the efficiency of the process and the quality of the results. Today's large designs often require multiple complex testbenches and several layers of integration. Waiting until the design is complete to begin the verification process increases the total development time. By making the verification process independent, verification can begin in parallel with the design process. Instead of waiting for testbenches to be created and tests to be written after the design is completed, a parallel verification process enables the development team to begin testing the design immediately after it has been developed.

Project managers have often described the design and verification of hardware or software as a "V" process, as shown in Figure 5. In a V process, the project moves from design at a higher system level to design at lower chip to the block levels. The design is then integrated, then verified and tested from the low block levels back up to the chip and system levels.

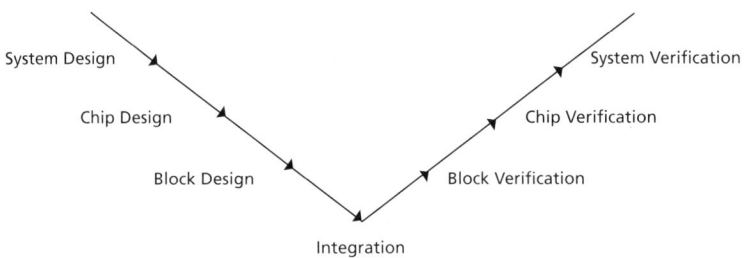

Figure 5. Basic V Design and Test Process

Advanced verification teams follow a modified V process. In the basic V process, the verification tasks include developing testbenches, writing tests, and running and debugging the tests. All of these tasks wait until integration is completed and follow in a linear fashion. Advanced verification teams accomplish the development of testbenches and tests in parallel with the design so that all that is left to do after integration is run the tests and debug. This modified V process is shown Figure 6.

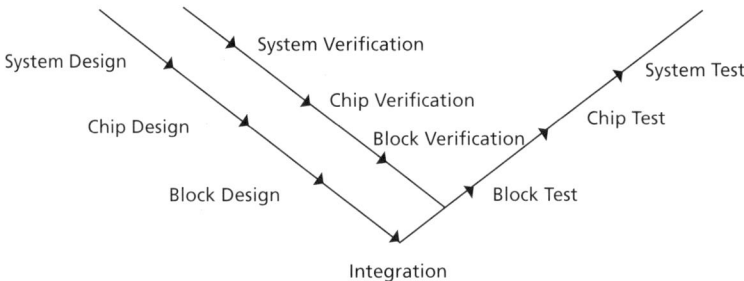

Figure 6. Modified V Design and Test Process

In addition to decreasing the total project time, a separate verification process improves the quality of the results. Verifying a design often requires a different mindset than implementation. When you implement a design, you concentrate on how the design should work. When you verify the design, you also concentrate on how the design might not work. It is common practice in many software development processes that the designer should never be responsible for testing one's own code, because if the designer made an incorrect assumption or interpreted a specification incorrectly when implementing the design, the same assumption or incorrect interpretation will be made in testing the design. Having an independent person verify the design decreases the likelihood that the same incorrect assumptions or interpretations are made in the verification. Engineers working solely on design verification provide an independent and focused approach that will find more bugs and find them faster.

Separating design and verification affects the organization and makeup of the development team. A complex project might require a large team of dedicated verification engineers who need a separate manager or lead to coordinate them and track their progress. Some teams might separate engineers into different groups focused on either design or verification. Other teams might have a single group of engineers and designate some as focused on design and some as focused on verification. Either way, the responsibilities for design and verification are separated, and the overall effect to the organization is the segmentation of verification engineers and the need for separate coordination of the verification process. Separating design and verification has affected the engineering community by providing a new area of specialization and a career path in functional verification.

COORDINATING VERIFICATION WITH OTHER DEVELOPMENT TASKS

For many teams, functional verification is considered independent of other development tasks, such as architecture, software development, or system design-in. In actuality, there is great overlap between each of these areas and functional verification. Architects do not simply create new systems by pulling ideas from thin air. They often need to prove that their ideas are feasible or choose between different ideas. The process of testing feasibility and comparing system responses is very similar to a functional verification process. In many systems today, a large amount of the functionality of the product is encompassed in the software that runs the design. Verifying that the software works with the hardware is an important part of the verification of these systems. System design-in is the process of taking a device and implementing it in a final product. System design-in teams need to verify that not only is the device functioning correctly but that it will also function correctly in the final product.

Development teams that do not understand and manage the coordination of functional verification with other development groups waste time and resources. A common thread among functional verification, architecture development, software development, and system design-in is the need for a representation of the design: architects to test new ideas or algorithms; software developers to test that the software works correctly; system design-in teams to verify that the system works together; and verification teams to verify the implementation. When each of these groups works independently, the work required to create and maintain these representations is duplicated. If teams coordinate or reuse the functional representations, they can greatly reduce development time and resources.

In the past, teams could wait for a complete architectural analysis to be completed before beginning implementation, and software development could wait until the hardware had been designed. But if teams want to meet today's reduced development schedules, they must perform these development processes in parallel. This means that design and functional verification begins before the architectural analysis is completed. Software development is done in parallel with hardware development, and the final system design-in begins before all the parts are fully tested and working in the lab. Starting these development processes earlier in the project creates new demands for functional verification teams. The verification team begins with less well-defined architectural specifications and needs to combine performance testing with functional testing. The other development teams need a verified representation of the design earlier in the project, so the verification team must prioritize their efforts and synchronize with other development teams.

Closer coordination between the verification team and the other development teams affects the communication and scheduling of the verification process. Functional verification is often not considered until after the project has begun and progressed for a period of time. The thinking of many design teams was that the verification team should not engage on a project until the architecture and specification were complete and the design was well underway. If the verification team is to coordinate with the other development teams, it needs to engage in the project sooner. Verification needs to understand the development of the architecture if it is to help with performance analysis and if specifications are limited. Verification managers or leads also need to understand the requirements for deliverables between each group and to schedule resources accordingly. Organizationally, the verification team plays more of a central role in the development process.

VERIFICATION AS A MULTITHREADED PROCESS

Verifying simple small designs is a serial process. The verification effort is usually accomplished with basic simulation done on the entire device all at once or in a few small parts. There are few milestones to track or parallel processes to coordinate. But as designs grow larger, teams need to break the designs into hierarchical pieces to be tested independently before being integrated. As designs become more complex, different methods and techniques need to be used to verify the design in different ways. Verifying a large complex design includes several parallel processes, multiple stages, and many dependencies to manage.

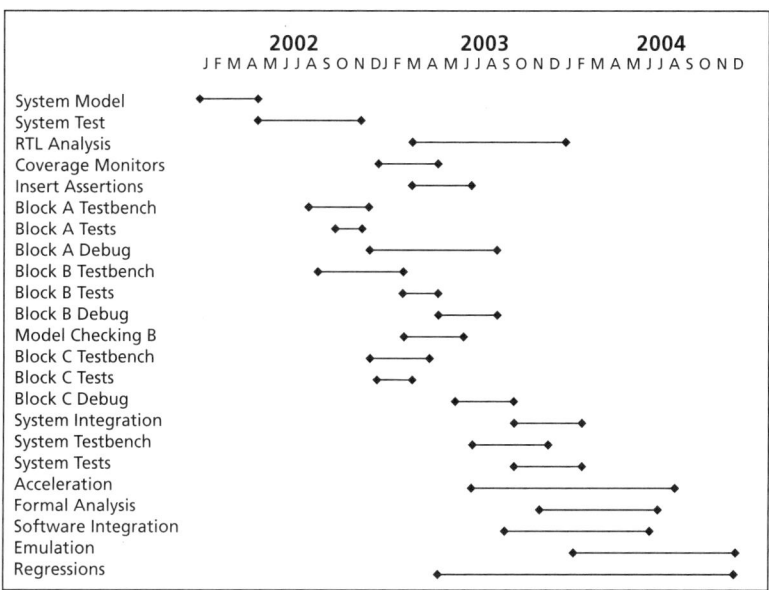

Figure 7. Verification Process with Many Tasks and Dependencies

If a complex verification project is not well managed, time is wasted and the quality of the results might be poor. A large verification project may consist of many environments to test the individual blocks, integrated subsystems, and final system. Each of these might have multiple dependencies, such as delivery of RTL, creation of testbench components, and creation of tests. If these dependencies are not managed well, engineers are stalled and unable to complete their tasks. The verification of a complex design can include multiple methods. The team might use simulation, static verification, emulation, and software coverification to verify different aspects of the same design. Managing the requirements of these methods so that each can be performed successfully improves the overall quality of the design.

A complex multithreaded verification process needs project management. A verification strategy must be developed early in the project to coordinate the multiple processes. A verification manager needs to be able to track the internal dependencies of the process as well as the status of the design and other development groups. Managing multiple dependencies requires a flexible verification team that is able to adapt to the inevitable events that shift priorities, resources, or schedules. The verification manager or lead needs to have project management skills and understand the entire development process to be successful.

VERIFICATION IS NOT 100 PERCENT

Verification of small designs relies on the ability to completely verify the design. In the most basic verification approach, the design is stimulated with every possible combination of stimulus and the responses are verified to be correct. As the design size and complexity grow, it quickly becomes impossible to test every possible stimulus combination. One approach is to limit the testing by running suites of stimulus that stimulate the exact scenarios the design is intended to operate under. This can only be done for designs that operate in a confined deterministic manner.

If the size and complexity reach a point where it is impossible to verify every possible scenario, the verification team needs to develop a new strategy to be able to meet the schedule and avoid bugs slipping through the process into silicon. Some verification teams prioritize the possible scenarios and make sure that the most important scenarios are tested first. This approach reduces the chances of a bug being found in an important area of the design. Some verification teams use automation and random testing to verify as much of the design as possible in the allotted time. This provides the widest possible coverage that a bug does not exist. Most verification teams use a combined approach of running the most important scenarios first and then using random testing to cover the widest areas later in the process. Whichever method is chosen, the most important thing is to choose a strategy that addresses the project's needs.

Being unable to verify the design completely affects the entire development process. Developing a new design is an exercise in risk management. The verification team cannot provide 100 percent confidence that the design is functionally correct. There is always some risk of a bug slipping through the process and being found in the lab or at the customer. The more effort and time applied to verifying a new design, the less risk of a missed bug. The management team needs to decide when enough verification has been done so that the risk of a missed bug affecting the design is acceptable. There are many factors that go into this decision, including time-to-market pressure, development costs, and quality concerns. The verification team can assist this process by tracking progress closely and providing metrics to help measure the risk.

VERIFICATION IS METHODOLOGY-BASED NOT TOOL-BASED

Today, standard verification is mostly done with simulation. As the design size and complexity grow along with the time-to-market pressures, development teams attempt to address perceived weaknesses in their verification with different approaches, such as assertion-based, transaction-based, or coverage-based verification. Each of these methods focuses on one or two individual verification issues, such as debug time or identifying coverage holes. Unfortunately, verification teams often find that when they base their verification on one method, it addresses the one issue but exacerbates others. Teams begin to add one tool or method after the other hoping to address their verification problems, but quickly find that due to incompatible tools and approaches they are spending more time addressing tool issues and less time verifying the design.

Verification tools associated with methods, such as simulation, assertions, coverage, or emulation, should not be the center of functional verification. These tools should be viewed as a collection of utilities to help reach your functional verification goals. A carpenter keeps a tool box full of different tools, but does not let the tools dictate how something is built or fixed. Instead, the carpenter figures out how to build or fix something and then chooses the right tools in the right sequence for accomplishing the task. In a similar manner, an advanced verification team needs to first understand what it is they are trying to accomplish, then develop a methodology using the right tools at the right time. In some projects, certain tools may be heavily utilized and, in others, they may be lightly utilized or not used at all. Functional verification becomes based more on the methodology used to reach the final goal than on the individual tools that may be used within the methodology.

Methodology-based verification places more emphasis on planning and strategy and less on techniques. Verification teams are often engaged late in the process and have no time to plan or develop a methodology. Thus, the methodology is developed on the fly, addressing whatever the most urgent issue is at the time. This leads to duplicated processes, wasted time, and poor results. When a team moves to methodology-based verification, they assess the needs of a project, plan how they will meet their goals, and develop the best approach for reaching their goals. Once a methodology is developed, the tools are brought together in a unified manner to address the project goals.

VERIFICATION DIFFERS FOR EACH DESIGN

The key to becoming a successful development team is not just the ability to develop one great product, it is the ability to develop multiple products. In the past, product life cycles were long enough that a development team could focus on one unique product at a time. In addition, most designs were very homogeneous—different functions, such as analog, algorithmic, digital, or radio frequency (RF), were developed independently on different devices. This meant that development and verification approaches from chip to chip were very similar. Today's product life cycles have decreased to the point that a development team may be finishing one design while they are working on the derivative design as well as the next generation. Designs are also now more heterogeneous with system on chips (SoC) containing multiple functions in the same device. The verification approach that worked for one chip may not work for the next.

Verification teams today need to evaluate each project before executing a strategy. Teams need to be able to adapt to different design types, different schedules, and different amounts of available resources. Two designs can be almost identical, but if the amount of resources or time available to verify is different, different strategies may be called for. This does not mean that every project requires a completely different strategy, but advanced verification teams need to be flexible to adapt to different circumstances while still keeping the same basic focus.

Verifying multiple designs has many implications. First, the verification team needs to accurately assess the needs of an individual team and plan a strategy to meet these goals. Teams cannot take a one-size-fits-all approach to a new project. A verification team also needs to develop a flexible and reusable verification methodology and supporting environment. Creating an environment that is flexible enough to support multiple projects with different needs takes more effort and time than creating an environment specific to one design. Management needs to understand the long-term benefits of developing a flexible environment and not push for short-term gains by cutting corners. The team needs to develop an environment based on open standards. Closed proprietary solutions might be the quickest solution for one project, but they are not expandable or adaptable. Open standards allow the environment to grow and adapt with changing needs.

There are many more characteristics of advanced functional verification, and there will always be new ones evolving. This book focuses on the techniques and processes currently used by teams throughout the industry. We will now look at how some of the successful teams employ the best practices of advanced functional verification to achieve their goals.

Chapter 4

Successful Verification
Managing time and resources using advanced functional verification

By studying a number of successful advanced verification teams, a set of common guiding principles emerge. These principles guide how teams perform the process of verification as well as manage their time and use their resources. Throughout this book we present best practices used by advanced verification teams that can all be traced back to these common guiding principles.

TIME MANAGEMENT

Time management is an important part of any complex process. With proper time management, verification teams can complete verification sooner or perform more verification in the allotted time.

Start Early

Every development project is unique and often requires new approaches for functional verification. Starting the process of verification early in the project enables the team to plan for new approaches and to adapt to changing environments. Starting early also allows verification to guide important early decisions, such as IP selection and feature support. As verification becomes a larger and larger portion of the development process, more decisions will need to be made to weigh the trade-offs and effects.

Functional verification requires preparation. If the verification team waits until the design has been implemented to begin, time is wasted developing and debugging the verification environment and tests. Verification teams need to be ready to test the implementation before it is received so that no time is lost.

Successful verification teams start by demonstrating the value of having verification knowledge early in the process. These teams become involved in the development and testing of system models used by architects and system designers. They also try to decouple the development of verification environments and tests from the implementation process so that they can be done in parallel. Of course, it is impossible for teams to engage in new projects early if they are still supporting older existing processes. Successful verification teams are careful to stage the roll-off of verification resources from past projects to synchronize with the early needs of new projects.

Remove Dependencies

The time it takes to complete a complex process like functional verification can be reduced by speeding the individual subtasks of the process or by removing the dependencies associated with the subtasks. Successful verification teams understand that time spent waiting for a deliverable from one task to start another task is wasted time. Removing dependencies not only decreases the amount of time to complete the overall project, it also uses the resources more efficiently throughout the project. Waiting for deliverables like HDL code or specifications leads to large spikes in resource utilization, followed by lulls as the resources wait for the next key deliverable.

Removing dependencies from external teams that are waiting for deliverables from the verification team reduces project time and improves the perception of the verification team. Implementation teams must wait for a bug-free design, and software teams often wait for functional models before beginning implementation. Staying off the critical path for the project should be an important goal of any verification team.

Successful verification teams remove dependencies in a number of ways. Many teams develop their own high-level system model to use in place of the HDL for developing tests and environments. The same system models can also be used as an executable specification, alleviating the need to wait for a functional specification. External dependencies can be met by providing the high-level model or an early prototype or emulation system to software and system design teams.

Focus on Total Verification Time

It is important to not lose focus on the big picture when concentrating on speeding up the individual processes and removing the dependencies in the verification process. Total verification time is the amount of time it takes from the start of the project until the design is declared functionally correct. Successful verification teams are aware of the larger project goals and adapt their plans and approaches to these goals. This allows these teams to use the appropriate processes and correct resources throughout the project to attain the project goals.

During a verification process, a team faces many trade-offs. When weighing trade-offs, you should always consider the effect on the total verification time and not simply the short-term milestone. When considering a new approach or technique, successful teams measure the run-time improvements, but also factor in the user time. Executing a test faster is only beneficial if it does not adversely affect the amount of time you have to spend setting up the test or interpreting the results.

Successful verification teams focus on finding the easy bugs in a design early in the fastest, most efficient manner. These teams understand the importance of finding the critical bugs that may result in redesign. They are aware that their most valuable commodity is designer time, because designers are often the only ones with the knowledge to debug a failure and make the appropriate fix. They want to utilize the designer's time in the most efficient manner possible.

RESOURCE USAGE

Managing resources within a verification process is not simply trying to optimize and do more verification with fewer resources. Resource management also includes building teams and environments that facilitate efficient high-quality verification.

Plan and Document

The planning process may be the most important and the most neglected part of functional verification. Creating a verification plan provides a process for developing the strategies and tactics that will be used on a project before it has begun. This planning provides a map for the team to use as guidance during the process and as a means to track progress. Successful teams often have mechanisms that automatically track test and development status to goals listed in the verification plan. Without a proper plan, verification teams are left in a reactionary position, often taking many wrong turns on their way to reaching the project goals.

Successful teams solidify the plans and processes by documenting them. Documentation provides a mechanism for communicating the internal and external expectations and deliverables to the group. This communication becomes more important as team members engage and disengage from the process. It enables existing team members and new team members to be on the same page throughout the project. Successful verification teams are also careful to not overload the process with documentation. These teams make concise documentation part of each deliverable, providing enough detail to track the process but not requiring unnecessary overhead.

Build a Team

The goal in building a successful verification team is to have a team whose total abilities are greater than the sum of its parts. Successful verification teams most often consist of individuals who have a common baseline knowledge of verification as well as specialized knowledge in specific areas. Each team member is familiar with basic test writing and debugging skills in addition to an in-depth knowledge in an area, such as software development, testbench development, scripting, or emulation.

Just as engineers learn the profession of verification through experience and mentoring, verification teams also are built through experience and benchmarking. As a team works together, it learns to utilize the individual skills to meet the goals in the most efficient manner. Successful teams also benchmark themselves against other teams and learn from the best practices of these teams.

Successful verification teams build a cohesive team by selecting the right members and keeping them together. These teams select members with a wide range of abilities and experience. The verification process has many complex tasks that require experienced individuals, and it has many basic tasks that can be performed by less-experienced members. Having a well-balanced team keeps everyone engaged in the process and provides a path for development. A trademark of many of the most successful teams is that they have worked together for many projects. Keeping a verification team together is often difficult, but the benefits are enormous.

Use Someone Skilled in Management

Managing a process as complex as verification requires experienced project management skills and a well-qualified leader. Project management requires planning, negotiation, and monitoring—skills not often found in the most technically talented individuals. Yet, in many organizations, the manager or lead of a verification group got their position by being the most technically skilled individual. Successful teams separate the need for competent project management from the need for competent technical leadership. A competent verification project manager provides value to the entire development team and uses verification experience to make the right trade-offs. The manager also provides a voice for the verification team to make sure that their concerns are heard.

Most successful advanced verification teams have competent project management as well as technical leadership. While these teams may not have specific roles or titles for these individuals, there are always one or two members who know what is going on and are coordinating the efforts of the team.

Successful teams develop their project plans and schedules in coordination with the overall project and do not let the other groups dictate their work. Organizations today understand the need for verification leaders and are designating lead or management roles for these individuals.

VERIFICATION PROCESSES

The verification process is made up of many smaller separate processes and techniques. Each of these smaller processes has its own unique tools and methods. Selecting these tools and techniques carefully results in the most efficient overall process.

Choose the Right Tool for the Job

Selecting the correct verification tool or method for a particular task is vital to maximizing the overall efficiency of the process. Using the correct tool at the correct time is a matter of understanding the needs of the task and the capabilities of the available tools. Many teams rely solely on the tools that they are most experienced with. Other teams try to always use the newest tool or technique to gain a competitive advantage. The truth is that no one approach can address all your problems. In some cases, using the tools that you are most familiar with is the most efficient course, and in other cases new tools are called for.

Selecting the correct tools can save a team more than just time and effort. Verification tools can be quite costly. Some teams believe it is cheaper to buy only a few types of verification tools rather than a wide range. Teams should factor in the cost of the total quantity of tools and not allow themselves to get tied into one approach or tool. Verification tool vendors have moved from providing individual point tools to offering platforms of tools. These platforms reduce the overall cost of verification and also give verification teams the flexibility to select the correct tool for their job.

Successful verification teams select the correct tool for a job by first evaluating the needs of the task. Simulation is often a trade-off between speed and control. Fast simulators, such as hardware simulators, have the best performance but often provide poor visibility and control for debug. Software simulators have excellent control and visibility but at slow speeds. Successful teams use lower performance simulation during the early stages of debug when bugs are plentiful and switch to hardware simulators when simulations require long run times. Many teams rely solely on simulation for their verification, but some bugs, such as clock domain crossing bugs, cannot be simulated easily. Successful verification teams realize these bugs can be detected faster with less effort with dedicated static tools. These are just a few of the possible trade-offs teams should make.

Choose the Right Information for the Job

Verification tasks require more than just tools; they require accurate and efficient use of design information. The most basic trade-off verification teams need to make is between the speed of a test and the fidelity of its results. Simulations can be performed at very high speeds but provide results that are not specific enough to be useful. Simulations can be performed with great amounts of detail but take a long time to complete. Choosing the right balance of detail and speed is vital for maximizing overall verification efficiency.

Design information is more than just the functional representation of the design. Input stimulus and output responses also characterize the design. Choosing the correct level of detail for this information facilitates the reuse of verification components within the testbench and eases test development. Choosing the correct level of detail for collecting coverage information and waveform data eases the collation and use of this data for debug.

Successful verification teams select the correct levels of information for the design representation and verification data at the beginning of the process. These teams might use a transaction-level model of the design to perform early performance and architectural analysis, which requires high speeds but only basic levels of detail. Later these teams use an RTL or gate-level representation of the design to verify the actual implementation details. Successful verification testbenches are often based on a common API that facilitates reusing components written with a common level of detail. Using these testbenches, along with common databases, facilitates the fastest, most efficient test and debug processes.

Automate

Automating the various tasks of the verification process reduces the time and effort the team spends on repetitive tasks. The verification process contains many individual tasks that might be repeated hundreds or thousands of times during a project. Automating these tasks might not speed up the individual task, but it does free up valuable resources to concentrate on other tasks. Resource loads can be balanced to utilize the available resources, such as computers and software licenses. Automation can also facilitate higher quality results. Human error can easily enter long repetitive tasks, such as running large test suites and collecting results. Removing the chance of human error assures more consistent reliable results.

Automation also helps in documenting the verification process. Proper scripting of the various verification tasks provides a verification record. These scripts can be used to understand the process if the project needs to be redone or modified at a later date.

Successful verification teams automate every chance they get. Any process that is performed more than a few times is automated. The most common automated verification tasks are building, compiling, and executing simulations. By the time a large project nears completion, a simulation regression environment can contain thousands of tests that need to be run and rerun with each change in the design or environment. Maintaining the integrity of the model as it nears tapeout becomes mandatory. As time moves on, parts of older projects are often incorporated into newer projects as IP. Automating the verification environment facilitates this process, making it easy for the new project to integrate and test the older code.

APPROACHES

Proper management of time, resources, and tools results in a highly efficient advanced verification process. The last group of principles encompasses the overall verification approaches that successful verification teams use.

Keep Verification Real

The closer the verification of a design reflects the design's real operating environment the better the quality of verification. It may seem obvious that verification teams should strive to create test environments and stimulus that reflect the real-world conditions the design will face. But many verification teams get caught up in developing complex testbenches and using complex formal techniques and lose the big picture. The goal of verification, and of the entire development process, is to produce a product that works as intended. This goal not only keeps the verification "real," it is also more efficient. It is impossible to verify everything in a large complex IC. Teams must prioritize, and the highest priority should be what is most important to the end product.

Another benefit of keeping verification real is that it becomes easier to understand. Test writers and debuggers can more easily comprehend their task if it can be related to a real application. Humans naturally think in terms of concrete concepts, such as real images or communications. Using verification environments that allow engineers to conceptualize what they are working on makes verification easier and more efficient.

Successful verification teams develop environments that reflect the real world whenever possible. These teams utilize input data such as images, program streams, or traffic flows taken directly from real-world applications. The teams also utilize tools, such as emulators that allow the design to be tested in a real-world environment. This extra verification not only verifies that the design works correctly, but also verifies that the testbench and models that were used are correct for future use. Finally, these teams write their environments so that tests can be written at a higher application level and results can be debugged at that same high level. Transactors or adapters are written to translate between these higher level data abstractions, such as an image or a packet, down to the signal level necessary for detailed verification.

Stress the Design

Completing a verification process can be a daunting challenge. Just completing the verification of the known functionality can take more time than is available. Unfortunately, many of the most difficult bugs are later found in areas of the design that were unknown to the verification team. Stressing the design outside of its normal operating parameters during the verification process can flush out these bugs and prove to be a very valuable exercise. Using processes that rely on random generation of stimulus can explore and stimulate the design in ways the verification team might have never considered. These techniques often find bugs that otherwise would not be found until the design were in the lab or at a customer site.

Stressing the design can provide more than just bugs. Often design teams believe that they know the bottlenecks in the design that are limiting the performance to an acceptable level. They may target these areas in later revisions to improve performance or to add new features. Stressing the design beyond the expected performance level can verify that these areas are indeed the bottlenecks and quantify their real effects. In many cases, unknown bottlenecks are discovered that would limit the actual benefits of future revisions of the product. In addition, by stressing the design with illegal data, the team can learn how the design will respond to these conditions. This information can be valuable for lab debug, fault detection, and error recovery efforts.

Successful verification teams concentrate on stressing the design outside of its normal operating parameters once they reach a stable level of verification. These teams use random stimulus generators within their testbenches to generate data, sequences of data, and timing relationships that may never have been considered. The teams also stimulate the design at data rates faster than required to oversubscribe the design and discover where performance bottlenecks exist. The teams apply illegal or unpredictable stimulus data to identify the reaction of the design to these possible error cases for future reference.

Chapter 5

Professional Verification
From second-class citizen to respected profession

Some might argue that functional verification is not a true profession because of its low visibility and secondary role in the development process. A closer look shows that today functional verification is an integral part of the development process, requiring professional skills and organization. The term professional has many meanings. Many people associate professional with being paid to do a job, such as a professional athlete or musician. One might also associate professional with the expectation of a high-level of quality. The Cambridge Dictionary defines professionalism as "the qualities connected with trained and skilled people," and a professional as "a person who has a job that requires skill, education or training." Professional verification encompasses all these meanings.

UNDERSTANDING PROFESSIONAL VERIFICATION

Professional verification is the practice of functional verification using advanced verification practices. The emergence of functional verification as a profession in its own right has come about slowly. In the not so distant past, verification was thought of as simply one part of the design process or as a stepping stone in the career path of an engineer. Today, if you walk into a company that develops advanced ICs or attend a conference on IC development, you are likely to find engineers with verification in their job title. Because the job of functional verification has become so difficult, it requires a new class of engineers: design verification professionals.

Professional verification assumes a certain standard of quality. Many development teams do just enough verification to get by. Their goal is to verify enough before the tapeout date and hope that nothing has slipped through the cracks. While this risk-management approach to verification is suitable for many efforts, one expects more from professional verification. A development team using professional verification expects first-pass success and a final design that meets all the functional requirements on schedule.

Professional functional verification requires specific skills, education, and training. It requires knowledge of advanced software development, digital logic design, failure analysis, design automation, and the understanding of complex protocols and applications. Functional verification is not a topic currently taught in engineering schools. Industry-training courses and available texts focus more on the tools or languages used in the verification process than on the actual practice of design verification. Engineers new to verification obtain the required skills by working with advanced verification engineers or by being thrown into the job and learning from the mistakes they make.

You may be wondering why this emphasis on professional verification, since in many companies verification is considered a lower class job. You see all the promotions, raises, and recognition going to the designers or architects of the ICs, and little if any credit given to the engineers doing verification. When times are hard, verification engineers are the first to lose their jobs and are treated as easily replaceable. The perception of many in management and design is that verification is for those engineers who are inexperienced or lack the skill for design or architecture. To get a better understanding of the need for professional verification, it is important to look at how verification impacts a company's success.

The lesson learned during the Internet boom and subsequent crash is that the most solid successful companies focus on profit. Companies that provide the most value to the customer for the least cost are successful. Every process is measured by how it impacts the company's profits. Functional verification is no different; its importance lies in how it improves value and reduces costs.

THE VALUE OF VERIFICATION

Several factors can affect the value of a product. Having a product to market first provides a direct value in not only the premium price that can be charged, but also in the early adoption of the product. Having more features in a product can also improve the value if it increases the market base or provides differentiation from competitors. There is also value in being able to quickly adapt to the changing requests of customers. Functional verification can add value to a product by positively affecting these factors.

Time to market is a common concern of many development teams. Verification is the most time-consuming task in developing ICs. If there is not a separate verification team, analysis has shown that designers actually spend more time doing verification than design. Reducing the amount of verification time through better practices reduces the time to market of a product. An analysis done by Cadence Design Systems showed that by using best verification practices a team can reduce the total time spent in verification by up to 50 percent. In addition, time is saved by doing quality verification. A functional error that is missed by the verification process and found in the lab, testing, or at a customer site can cause months or even years of delay in getting the product to market.

Deciding which features to include in a product is not determined by the time or difficulty in implementing them, but in the time and cost it takes to verify and test them. In many cases, adding a new feature to an IC is as easy as adding a mode register and some random logic. But if you are doing basic brute force verification, this simple change can cause the verification effort to grow exponentially. Advanced verification based on adaptability and reusability can more easily handle and limit the effects of large feature sets. Efficient advanced verification allows for the verification of more features in the same amount of time, thereby increasing the value of the product.

When economic times are hard, adapting to rapidly changing customer needs becomes a must. If you have a rigid development environment, including an inflexible verification environment, it is difficult, if not impossible, to meet your customers' requests. Having an adaptable, reusable, and efficient verification environment enables you to respond to customers' needs and provide greater value.

What is the value of getting your products to market sooner, with more features, higher quality, and meeting your customers' needs? Professional verification can help you reach these goals. But what is the cost of achieving this value?

THE COST OF ADVANCED VERIFICATION

The cost of the raw materials to manufacture an IC is minimal; the real costs are in development. First, there is the labor cost of designing, verifying, and testing the device. Then there is the investment in resources, such as computers and software licenses. There is also the cost of producing prototypes before the manufacturing process is started. Finally, there is the cost of lost revenue when the market window for a new product is missed.

Many companies attempt to cut labor costs by either hiring the cheapest engineers possible or overloading their designers with additional verification tasks. The belief is that a team of inexperienced engineers or contractors is as productive as a smaller team of more experienced advanced verification engineers. This approach might make sense for one project or for projects of limited size and complexity. But if the cost of hiring, retraining, and redeveloping verification environments for each new project is compared to developing an advanced, automated, reusable, and flexible verification environment over the period of several projects, the advanced approach will almost always be more cost beneficial.

There is an old saying that you cannot put a price on quality. Professional verification assumes a high standard of quality, and some may wonder if the cost of that quality is worth it. The most measurable form of quality in functional verification is missed bugs. A bug that makes it through the verification process comes at great cost to the development team. A bug found after tapeout may require a respin or metal fix for the device, costing engineering dollars as well as lost time. A bug that makes it into the lab or validation process risks not only a respin of the device but the costs of time and dollars in reverifying and revalidating the IC or system. A bug that makes it to the customer can lead to additional engineering costs, but can also cause a loss of business, which is far worse.

Respins, metal fixes, reverification, revalidation, and customer bugs are common experiences for all IC development teams today. While the additional cost in time and resources for advanced verification might seem like an unnecessary investment, the potential savings in time and money provides a far greater return than a low-quality job.

VERIFICATION: SECOND-CLASS CITIZEN

If verification plays such an important role in the development process and can be shown to improve a company's bottom line, why do most companies treat it so poorly? Some of this poor perception has to do with the evolution of verification. For many years, teams could thoroughly verify their designs with simple techniques. The challenge of designing and implementing a simple design far outweighed the effort required to verify the design. Thus, verification was viewed as a straightforward process that was not as important as other areas in the development process.

As the size and complexity of designs grew, the verification problem began to explode. The more enlightened teams understood that verification was quickly becoming one of the most important parts of the development process and required attention and respect. But other teams still felt that they could make do with verification as they had in the past by patching the holes in their process. The complacency toward verification is like the story of the frog. If you put a frog into a pan of boiling water, it will immediately react and jump out. But if you put it in cold water and slowly turn the temperature up, it will sit and boil to death. Many development teams have been lulled by their past successes without paying attention to the changing needs of verification and are finding themselves in boiling water, not knowing that it is time to jump.

The Perception of Verification

One challenge any group responsible for testing or assuring quality has is the perception by the larger organization that they are holding up progress. The most visible task the verification team does is find bugs, which most often identify mistakes made by other team members. While a verification engineer may be proud or even excited to have discovered a difficult bug in the design, the rest of the organization often views this as bad news or a setback. It is human nature to be embarrassed by mistakes, and most organizations plan for success. Verification teams expose the mistakes of others and are the first to identify when a project is off track. This leads to the negative perception that verification is a necessary evil rather than seeing it as a positive factor in the development process.

Organizational issues often contribute to the negative perception of functional verification. Verification is often considered a training ground for new engineers or a dumping ground for poor engineers. Companies want their most senior and most qualified people to be in the role of designer or implementer. New hires are often first put in verification to "prove themselves" or to "pay their dues" before advancing to other development positions. Engineers who do not perform well in other roles are often moved into verification where they can "do less damage."

With most of the talent in the organization residing outside the verification team, it is likely that the engineers promoted to management are not from verification backgrounds. If you look at most engineering or project managers today, few have verification backgrounds. This lack of knowledge within management leads to poor decisions regarding the importance of verification.

Verification Training

Formal verification training is hardly available. Few, if any, verification schools offer functional verification as part of their curriculum. The reason most often cited is the lack of "pure research" areas in functional verification today. Only recently have periodicals begun to address verification topics or have organized verification conferences. Advanced verification is most often learned through experience and working with dedicated professionals. This has led to small pockets of functional verification expertise throughout the industry, but very little organization. Areas such as silicon process technology and design automation have thrived because of organized groups and consortiums, which do not exist for functional verification.

CHANGING CURRENT PERCEPTIONS

While many believe that functional verification will always be an after-thought in the development process, there are actions that can be taken to advance the perception and profession of functional verification.

Develop the Profession of Verification

Verification engineers need to become more visible in the industry. Verification leaders should work within their companies to promote the belief that verification is an important part of the business of developing ICs and requires dedicated professionals. Career paths should be established for talented engineers to remain in verification. Some companies are implementing cross-training by requiring engineers to be well-trained in verification as well as design and implementation. Project managers and future business leaders need to have some background in verification as well.

Externally, verification engineers need to become more visible in publications, organizations, and in schools. Verification engineers should stop lamenting the lack of training or relevant articles and start generating demand. Once publishers and training institutions understand the true size of the verification market, they will begin to address it.

Set Standards for Excellence

Verification engineers cannot expect their organization's view of their value to improve until they set some real standards for the value they provide. Expectations of what is good or excellent verification vary widely from team to team and company to company. Some companies believe they have a world class verification team but still routinely have multiple spins of their chips and find bugs long after verification is done. There is simply no standard for excellence in verification.

Advanced verification teams need to set the standard for what is expected of verification. First-pass success should be an accepted standard, not a wish. Verification teams should be able to provide accurate metrics on the quality of the design and the progress of the verification process. Verification teams need to attain these standards on schedule repeatedly, so the difference between advanced verification and ad hoc "just get it done" approaches can be clearly seen.

Train and Develop Leaders

The process of verification is different from other processes in the development of an IC. While most processes are driven by building or creating something, the verification process is driven by integrating and testing an existing design. The verification process has many interdependencies and requires a balance of coordination with the development of the design and the independent development of test environments for the design. Teams who attempt to perform advanced verification without a detailed plan and without leaders to drive the plan will most likely fail.

Share Best Practices

Verification teams need to work together to develop and share best practices. Consultants and engineers who have moved from company to company have come to understand that while some verification teams have strengths in some areas, no team is strong in all areas. Every team can benefit from an understanding of best practices in a wide range of verification topics. Many verification teams believe that their processes and approaches to verification are a competitive advantage, and that sharing this information with other teams would negate that advantage. While no team should willingly give their competition an advantage, best practices can be shared in open forums in a general manner. Describing a best practice and implementing it are two different things. In the next section of this book, we will now look at a unified verification methodology based on the best practices of experienced advanced verification teams.

SECTION 2
THE UNIFIED VERIFICATION METHODOLOGY

Chapter 6

The Unified Verification Methodology
A new approach to verification

The previous section described the issues in advanced functional verification today and detailed the need for a verification methodology that removes fragmentation and improves the speed and efficiency of the process. This section presents the Unified Verification Methodology (UVM) developed by Cadence Design Systems, which addresses this need. In fall 2002, Cadence Design Systems assembled a group of experienced verification engineers with varying industry backgrounds to create a unified verification methodology. The team was not limited to using a specific set of vendor tools or practices. The only constraints were that all the methods explored were in use today and that the resulting methodology dramatically increased the speed and efficiency of the verification process.

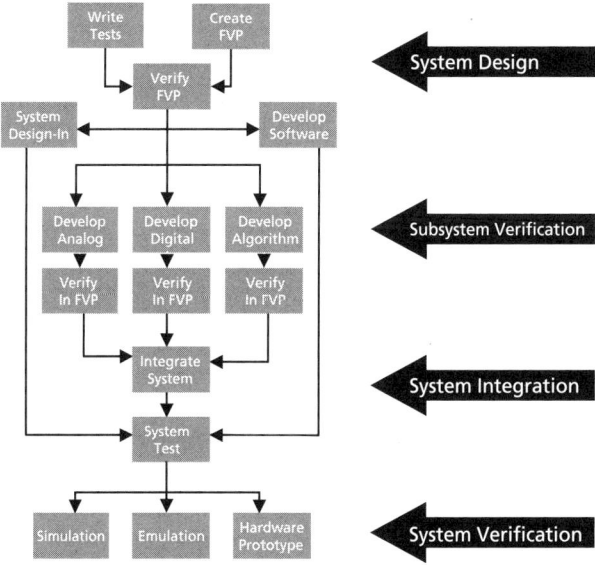

Figure 8. The Cadence Unified Verification Methodology

The first version of the Cadence UVM was released in February 2003, and since then advanced verification teams have used the UVM as a blueprint for developing their own unified verification methodology. The UVM program continues at Cadence as the team identifies new practices and refines the methodology based on real customer experiences. This chapter describes the key concepts of the UVM.

WHAT MAKES A METHODOLOGY UNIFIED?

Teams have always worked at bringing together the different tools they use into an integrated solution. A unified verification methodology, however, brings together the tools and technologies used in various tasks as well as unifies the processes and the people associated with the verification process. In a unified verification methodology, the tools work together in a coordinated fashion. Different tools run in parallel so that processes are not duplicated for each tool. The collection of data is shared among tools so that duplicate capture mechanisms are not required. The results are then presented in a common format or environment for the user to easily understand.

Traditional methodologies are often broken into individual stages, with rigid entrance and exit criteria for each stage. These criteria often create dependencies between processes, making it difficult or impossible to move back and repeat a previous stage. The stages of a unified verification methodology are tightly coupled and less rigidly defined, so it is easy to move forward or backward in the process without losing time or data. The team can move as far down the process as necessary with the information available. As more detailed or updated information becomes available, the process can be easily repeated. This continuous refinement enables the team to move forward without dependencies and adapt as new information becomes available.

Perhaps the most important aspect of a unified methodology is that it facilitates communication. The number of people involved in developing a complex system, coupled with the amount of data associated with the development, can lead to gross inefficiencies and functional bugs if not handled in a coordinated manner. A unified verification methodology provides the processes and mechanisms for facilitating efficient high-quality communication across the development team so that engineers have the correct and most up-to-date information they need to accomplish their jobs.

IMPROVING SPEED AND EFFICIENCY

The rate at which a task or event is performed is critical for a complex process like IC development, which consists of many different tasks. Improving speed involves more than just using faster CPUs or faster simulators. Rather it takes organizing the verification tasks into a unified methodology to provide the mechanisms and infrastructure for applying the needed performance gains. For example, the major focus in most IC verification today is simulation, which is done at many different stages of the verification process at many different levels of abstraction. Improving the overall speed of the simulation process is a major factor in increasing the speed of a verification methodology. Increasing the speed of developing testbenches and finding and fixing functional bugs also contributes to improving verification speed.

In today's resource-limited environment, we also need to improve efficiency. The verification process is constrained by the amount of human resources, compute resources, software costs, and time available. It is important to find an optimal balance between the time and energy needed to complete a task. Removing dependencies and ensuring that each phase in the process maximizes the reuse of previous stages improves the efficiency of the verification process.

The Cadence UVM is focused on dramatically improving the speed and efficiency of the verification process by unifying the different process stages and design domains to eliminate fragmentation in the verification process. The methodology is based on best practices used by advanced functional verification teams today. The best practices come together in the UVM to provide a blueprint for advanced functional verification. The methodology covers all process stages from system design to system design-in and addresses all the design domains found in today's complex System-on-Chips (SoC), including analog, digital, and algorithmic digital designs.

KEY CONCEPTS

Before we dive into the UVM, it is important to first understand several key concepts that run throughout the methodology.

Functional Virtual Prototype

A functional virtual prototype (FVP) is a golden functional representation of the complete design and its testbench. For many years, system models have been used in the development of ICs. System architects and designers have used system models for architectural analysis and to perform early functional trade-offs. Software developers have used them to run and debug their hardware-dependent software before it reaches the lab. The difference between these past uses of a system model and an FVP is that an FVP's primary focus is the functional verification of the design. An FVP unifies the use of a system model for software, architectural analysis, and functional verification to reduce the work required to develop this model. The central focus of an FVP is functional verification; it is the unifying vehicle for a unified verification methodology.

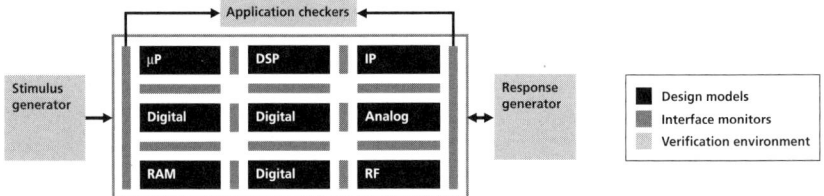

Figure 9.   The Functional Virtual Prototype

The FVP consists of functional models, interface monitors, and testbench components. The FVP is segmented into various functional sub-blocks in a manner similar to the implementation. Each sub-block is modeled to reflect the functionality of that block accurately. A block may first be modeled at a high level of abstraction based on an architectural specification or a behavioral model. The first version of the FVP can serve as an executable specification. Development teams can use this version as a source for golden models to verify the implementation of the functional blocks. As the blocks are implemented, they can be substituted back into the FVP, creating a mixed-level model for integration testing. The FVP is a continuously changing and evolving prototype. At any time, the FVP can consist of a variety of different models at different levels of abstraction.

It is important to differentiate between a model's accuracy and fidelity. An accurate model always returns a correct answer, although the answer might entail a broad range of responses. The fidelity of a model describes how closely the model represents the responses of the final implementation. An FVP must always be accurate in providing an answer or range. Its fidelity might change, depending on the knowledge available and the intended use.

The FVP also contains interface monitors that are located at the primary interfaces of the design and between each functional sub-block. The interfaces must support the transferring of information and data between each sub-block at different levels of detail or abstraction. High-level models may transport data as a data structure or as transactions. Implementation-level models may transport data at signal or bus level. Independent of the level of information transferred, information monitors verify the correctness of the data and the protocol of the transfer to ensure that correct data is being sent and received by each block.

The final parts of the FVP are the testbench components. Usually, a system model is considered independent from the verification environment. The testbench is included in an FVP to provide a common reference to the test suite and infrastructure during each step in the verification process. At any point in the development process, the FVP test suite can be run to verify that the design still meets the original system goals.

A question commonly asked about an FVP is how do you verify that is it correct? If the golden model is to be used to verify the implementation, an error in the golden model will result in an error in the implementation. The FVP should be verified in a similar manner as the final silicon will be verified in the lab. The test suite should verify that the design meets the specified level of functions, features, and performance. The test suite and test components of the FVP are very application-specific. Providing the testbench and application-level test suite ensures that the FVP continues to meet the original product goals.

The FVP is used throughout the UVM. As the design is being architected, the functional models are developed at a high level and tested within the FVP. Once the architecture is complete, the FVP is provided to each team developing a sub-block. These teams can reuse the model and testbench from the FVP to verify the implementation created by the sub-block design team. After the sub-block is verified by the individual development teams, the functional model for the sub-block is replaced in the FVP by the implementation of the sub-block. The test suite is run on this mixed-level FVP to verify that the implementation meets the application-level goals and to verify the integration of the sub-block. Each sub-block is replaced in the FVP until all have passed. Finally, all the models are replaced within the FVP, and final application-level testing is performed on the implementation-level FVP.

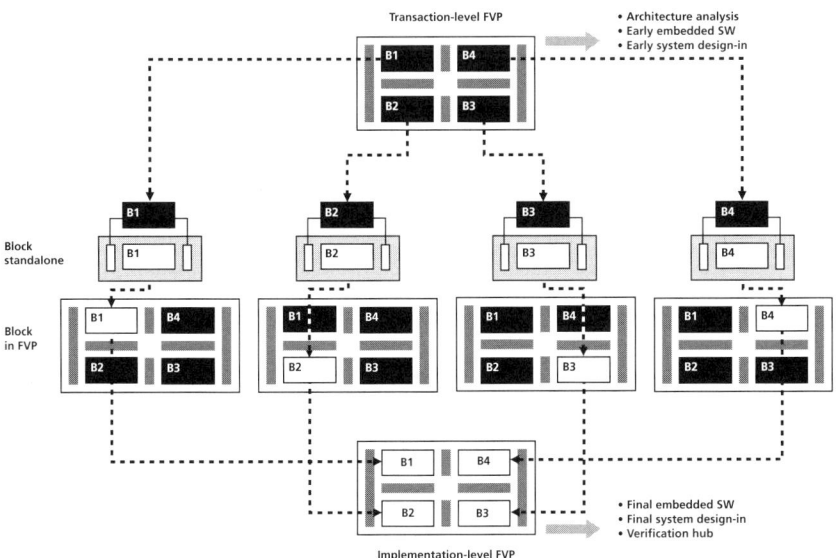

Figure 10. Using an FVP in the UVM

The FVP serves several critical roles in the methodology:

- Unambiguous executable specification
- Fast executable model for early embedded software development
- Early handoff vehicle to system development teams
- Reference for defining transaction coverage requirements
- The source for subsystem-level golden reference models
- Golden top-level verification environment and integration vehicle

Transaction-Level Verification

A dramatic improvement in design productivity occurred in the 1990s as designers moved from working at the Boolean gate level to RTL. Moving to RTL enabled designers to operate at a functional level that was more intuitive than simple gates, since designers think in terms of finite state machines, arbiters, and memory elements. However, as design sizes have increased and more functionality is placed in a single design, the verification process also needs to move up a level in abstraction from RTL to the transaction level. Verification engineers operate at an application level, where the concerns are complex data formats, algorithms, and protocols. To be productive, verification engineers need to think and work at the more intuitive level of packets, instructions, or images, and not at bus-cycle levels.

A transaction is the representation of a collection of signals and transitions between two blocks in a form that is easily understood by the test writer and debugger. A transaction can be as simple as a single data write operation or sequences that can be linked together to represent a complex transaction, such as transferring an IP packet. Figure 11 shows a simple data transfer (represented as B) linked together to form a frame, further linked together to form a packet and finally a session.

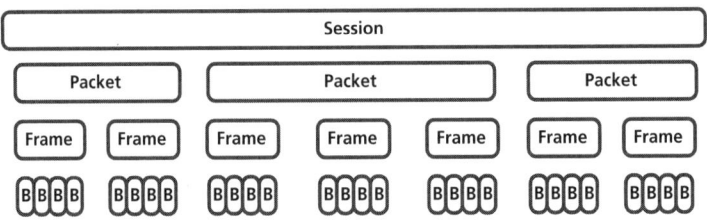

Figure 11.   Example of a Transaction Taxonomy

Transactions are used throughout the UVM. A transaction taxonomy that specifies the layers of transactions, from the simplest building block to the most complex protocol, is created early in the development process. Having a common transaction reference enables testbench elements and analysis tools to be reused. The following table shows the transaction taxonomy for Figure 11.

Table 1.   Transaction Taxonomy

Level	Data Unit	Operations	Fields
Interface	Byte	Send, Receive, Gap	Bits
Unit	Frame	Assemble, Segment, Address, Switch	Preamble, Data, FCS
Feature	Packet	Encapsulate, Retry, Ack, Route	Header, Address, Data
Application	Session	Initiate, Transmit, Complete	Streams
Interface	Byte	Send, Receive, Gap	Bits

Using transactions in the UVM improves the speed and efficiency of the verification process in several ways:

- Provides increased simulation performance for the transaction-based FVP

- Allows the test writer to create tests in a more efficient manner by removing the details of low-level signaling

- Simplifies the debug process by presenting information to the engineer in a manner that is easy to interpret

- Provides increased simulation performance for hardware-based acceleration

- Allows easy collection and analysis of interface-level coverage information

Unified Test Environment

The core of any verification methodology is the test strategy. Quite often teams are encouraged to base their entire test strategy on the latest tool or technique, regardless if it is the most appropriate. Instead, the UVM unifies the test environment by utilizing the strength of the appropriate tool or technique for the specific task in a common infrastructure. The goal is to create a unified test environment that is highly effective and efficient. We will briefly touch on some of the components in this environment, and go into greater detail in subsequent sections.

Assertions

Assertions are created in the UVM whenever design or architecture information is captured. You can use verification tools to verify the assertions
either in a dynamic manner using simulation or in a static manner with formal
mathematical techniques. These assertions are then used throughout the verification process to verify the design efficiently. The UVM uses three types of
assertions:

- Architectural assertions prove architectural properties, such as fairness and deadlocks.

- Interface assertions check the protocol of interfaces between blocks.

- Structural assertions verify low-level internal structures within an
  implementation, such as FIFO overflow or incorrect FSM transitions.

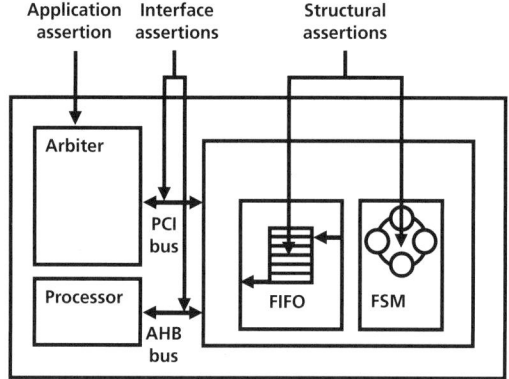

Figure 12.   Example of an Assertion

Assertions improve the speed and efficiency of the verification process in
several ways:

- Speed the time of locating difficult bugs by identifying where in a
  design the bug first appears

- Automate the instrumentation of monitors and checkers within the
  design

- Quickly identify when stimulus or interfacing blocks are not behaving as the implementer intended

- Identify bugs that did not propagate to checkers or monitors

- Detect protocol violations even if they do not cause a functional error
- Provide feedback to stimulus generators to modify their operations as the test progresses

Coverage

Coverage information improves efficiency by identifying areas of the design that have not been stimulated, tests that are not testing what they were intended to test, and functionality that is incorrect. Coverage cannot determine when the design has been completely verified, but it can indicate areas for more concentration.

The UVM uses four types of coverage information:

- Application coverage identifies whether specific high-level features of the design have been stimulated, such as automatic data retries.
- Interface coverage identifies whether sequences of stimulus and responses at the interfaces of the design under verification (DUV) have been verified.
- Structural coverage monitors the operation of low-level structural elements, such as FIFOs and FSMs, to identify which parts of the implementation have been verified.
- Code coverage identifies which areas of code have been stimulated.

Table 2.   UVM Coverage Types

Coverage Type	When Defined	When Measured	Examples
Application	Architecture Definition	System Modeling and System Verification	Auto Retry, Cache Hit
Interface	Implementation Definition	After Block Tests and Subsystem Tests	Packet Types, Instruction Sequences
Structural	Micro-Architecture Definition	After Block Tests	FSM States and Arcs, FIFO Thresholds
Code	Coding	After Block Tests	Statement, Expression

Coverage improves the speed and efficiency of the verification process in several ways:

- Identifies areas within the design that have not been tested
- Guides testing to the most important areas of the design
- Ensures intended functionality has been tested
- Used to select the most efficient and effective suite of regression tests

Hardware Acceleration

Accelerating the speed of simulation allows more testing to be completed in a shorter period of time. The key to using simulation acceleration in the UVM is choosing the most efficient method for the test you are running. As the verification process moves from short unit-level tests to longer subsystem-based tests, it is important to monitor the performance and debug times required for the tests. Hardware acceleration should be used once the test process has reached a stage where the run time is the dominant performance factor. Acceleration-on-demand enables the user to switch from simulation-based testing to hardware-accelerated testing using the same environment for development and debugging. An example of this process is shown in Figure 13.

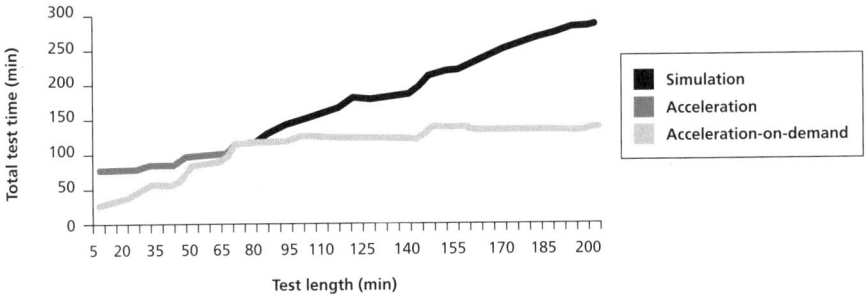

Figure 13. Example of Acceleration

This example is for a small block design with a non-synthesizeable test-bench. The standard run time for tests of an average-size design block is shown on the X axis. The Y axis shows the total test time, including compile time, run time, and debug time. Figure 13 shows that for short tests simulation is the fastest method. Once the test length reaches around 70 minutes, acceleration is the fastest. This is the point where acceleration-on-demand is effective. Each team should do a similar analysis, taking design size, test-bench performance, and debug times into account to determine when to accelerate a simulation.

Acceleration also speeds regression testing. Often small changes to the design or test environment can have unwanted and unknown effects on other parts of the system. Development teams set up a regression environment to verify that changes in the design or environment have not caused unwanted effects. It is important for development teams to receive confidence in their changes in the shortest possible turnaround time. Regression testing is often performed on large server farms running jobs in parallel. The time for completing the regression is dependant on the number of tests, length of the individual tests, the number of servers in the farm, and the simulation speed. When the number of tests and length of tests causes the total regression time to exceed the acceptable turnaround time, acceleration should be used. Acceleration reduces the run time of longer tests and allows large groups of shorter tests to be run in less total time.

Finally, acceleration provides the necessary speed improvements to facilitate system verification. Acceleration can be used in a simulation-based system verification environment to simulate large system integrations with embedded software. Acceleration is also at the heart of emulation-based system verification. Accelerating the design in an emulation environment connects the design to real-world stimulus and instrumentation while running and developing the system software.

METHODOLOGY OVERVIEW

Before diving into the details of the methodology, it is important to understand the overall flow from a high level. Perhaps the best way to describe the methodology is to compare it to the basic methodology used by teams today. Most development today begins with a group of architects or system designers working together to define the product at a high level. Simple models or spreadsheets might be used for reviewing design trade-offs and partitioning. This effort often results in a written architecture specification, which is given to an implementation team or group of teams to develop. In many cases, the implementation teams create an implementation or micro-architecture specification detailing what they intend to implement. The designers begin implementing individual blocks of the design in parallel. At some point, verification engineers may help each block team verify their block.

As each block is completed, it is integrated with other blocks, until the complete device is assembled. At this point, integration testing is done at the device level to verify that the blocks work together correctly. When the device is functionally correct, feature testing and performance testing verify that what was created meets the original product-level goals. If errors are found in integration testing or during performance or feature testing, the design is sent back to the development teams for fixing or redesign. After the design has passed functional, feature, and performance testing, it can be given to the back-end teams, software developers, and system design and verification teams.

With this methodology, testing is uncoordinated and occurs late in the development process. Once the architecture specification is agreed upon, each team works independently on their portion of the design. There is no mechanism to verify design assumptions, that changes still meet the original goals, or that these assumptions and changes are in sync with what other groups are expecting. The individual teams must wait until integration testing to discover problems due to incorrect, out-of-date, or incomplete specifications. The teams must also wait until final feature and performance testing to know whether the individual parts they created meet the goals of the device as a whole. These delays often result in bugs found late in the development process, causing individual blocks to be redesigned while the project is at a standstill.

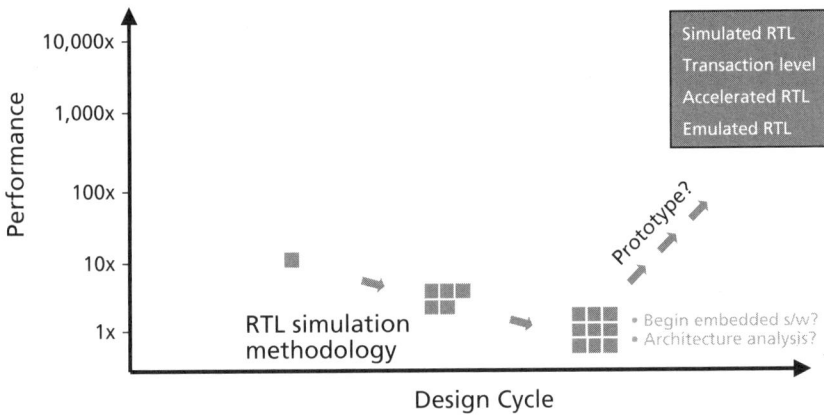

Figure 14.   Standard Verification Methodology

Another significant issue is the lack of performance at key times. In the early stages of development, simulation performance is appropriate for verifying individual blocks with isolated tests. However, as blocks are integrated together and test suites become more complex, simulation performance of the full device can degrade to less than 10 percent of original levels. Yet this is when integration testing, performance analysis, and feature testing are done, and where software development needs to begin. All of these tasks require high performance simulation, but this is where the worst performance is found.

Teams attempt to overcome this performance bottleneck by using a hardware emulator or an FPGA. Unfortunately, without proper preparation, converting to an emulator or FPGA can be a daunting task, requiring many resources and weeks or months of work. Even when using an emulator or FPGA is planned from the beginning of the project, it is difficult to predict the final results. The solution for many teams to this performance bottleneck is to spin early silicon and hope the device is functional enough to complete testing and start software development.

Advantages of the UVM

The UVM begins with developing a transaction-level FVP early in the development process. The verification team develops the FVP as the architecture is being finalized and the micro-architecture is beginning. A common mantra of many project experts is to begin with the end in mind. The verification team begins by developing a test plan for the final system based on the functional, feature, and performance requirements specified by the architects and system designers. The verification team is engaged in the project earlier than before to develop the FVP and work with the architects and system designers to develop the application-level test environment. This enables them to get up to speed sooner, help with architectural and performance analysis, and provide input into the testing strategy before the design is finalized.

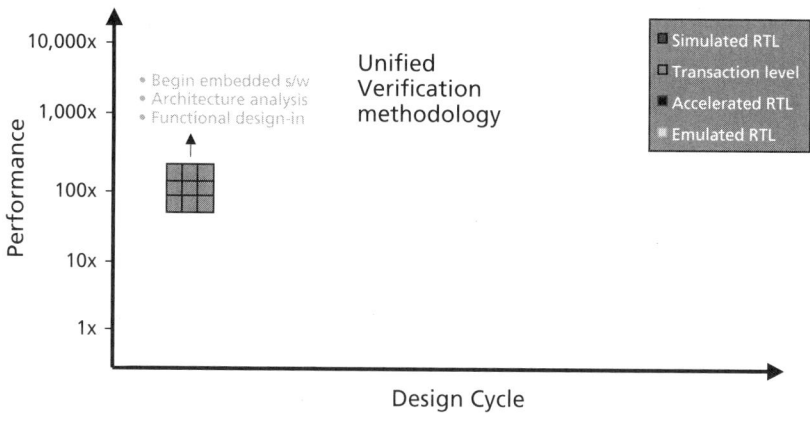

Figure 15. Creating the FVP

Once the FVP is completed, it is distributed to the various groups participating in the project. The system designers use the FVP to model the device in the context of the larger system. The software developers use the FVP to begin developing low-level hardware-dependent software, such as device drivers. The implementation teams use the FVP as an executable specification to develop the individual blocks that make up the final system. The block-level implementation teams use the FVP model as a reference model to verify that the responses of the implementation match the responses of the model.

The design is also verified by substituting the implementation for the model in the FVP and rerunning the FVP test suite. If changes are required to the block as implementation occurs, the changes can be verified within the context of the FVP and then propagated out to the other development teams. Verifying the implementation first against the model from the FVP and then in place of the model in the FVP verifies early in the development process that the design meets the specified requirements and works correctly with other blocks.

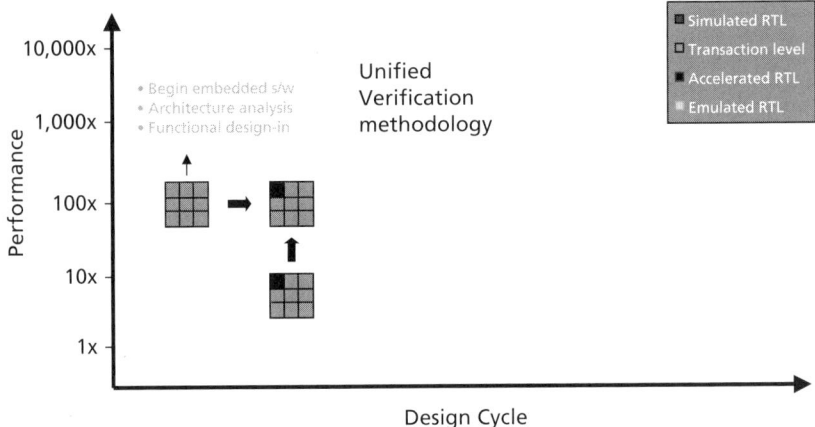

Figure 16. Block-Level Verification

As blocks increase in size and testing becomes more complex, simulation performance degrades. The UVM overcomes this performance degradation with hardware acceleration. With the UVM, each block is moved into a hardware accelerator to provide the required performance. A hardware accelerator allows the testbench to run at its original speed on a workstation while the design is mapped into hardware for acceleration. Each block is tested and added to the hardware accelerator as it is completed, first alone and then with other accelerated blocks.

After this, integration and final testing should be straight forward, since each block has already been verified to meet performance and functional goals within the FVP and has been verified to work with the other models within the FVP. In addition, the entire device has been hardware-accelerated, so the necessary performance is available. The final step is to include all the implementation blocks and rerun the FVP test suite on the entire design. Certain tasks, such as software integration and system verification, might still require added performance. Given that the design has already been run through a hardware accelerator, the process of moving to an emulator is one of disconnecting the testbench and hooking the design to an in-circuit environment.

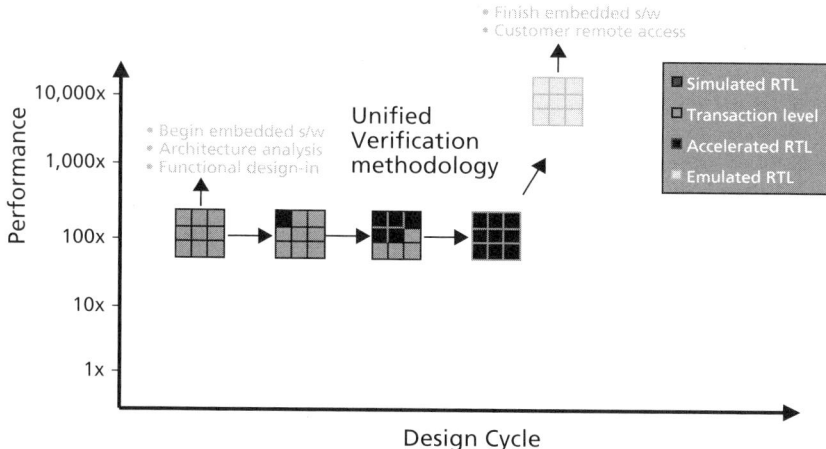

Figure 17. System Verification

The UVM provides advantages over a standard RTL methodology in both speed and efficiency. Through a combination of transaction-level modeling and hardware acceleration, the UVM can provide a 100X performance improvement at the chip level throughout the verification process. Early integration and software test allows you to find issues sooner and introduces parallelism into the development process to improve overall efficiency.

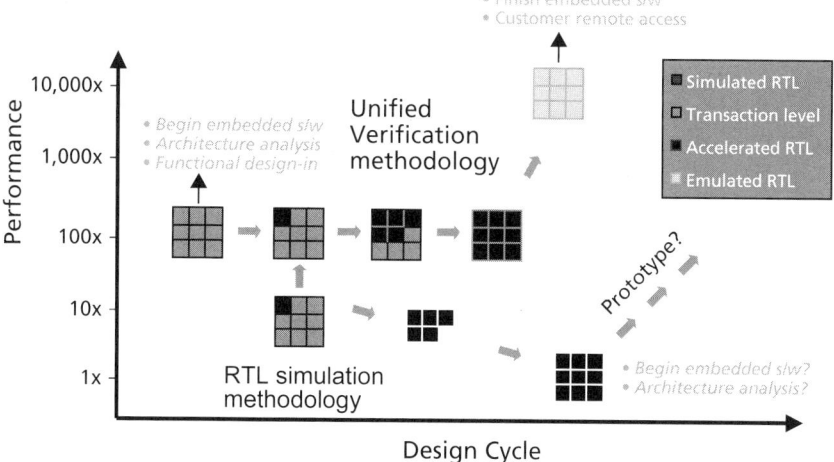

Figure 18. Comparing UVM with Standard RTL Methodology

This high-level description of the UVM has left out many of the important details on how to accomplish this methodology. The following chapters of this section describe more thoroughly how the UVM can be used for your design.

Chapter 7

UVM System-Level Design
Creating an FVP

The first stage of the Unified Verification Methodology is system-level design. During this stage, the product is defined, architectural measurements and trade-offs are made, and detailed specifications are created. In many methodologies, verification plays only a limited role. In the UVM, the verification effort begins early with the development and verification of the functional virtual prototype. In the UVM, the first stage is vital, because it sets the groundwork for unification throughout the rest of the methodology.

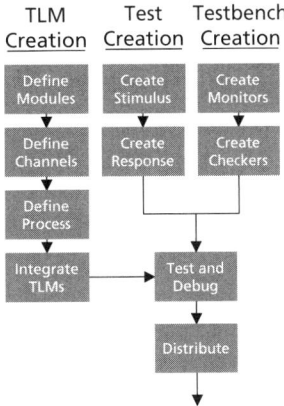

Figure 19. UVM System-Level Design Flow

This chapter discusses the differences between an FVP and a traditional system model, the costs and benefits of using an FVP, and creating and verifying it.

FROM WHITE BOARD TO FVP

Projects are started in many different ways. Some projects begin as an original idea developed from a blank sheet of paper. Others take an old idea and create a new implementation. Some simply modify and improve an existing design. But no matter how the project begins, most follow a similar progression from the initial idea to the final implementation. They only differ in where they start in the progression.

Most projects go through three basic stages: idea generation, product definition, and implementation. The idea generation stage is the most dynamic and where the steps taken differ from project to project and team to team. The result of the idea stage is often a document that specifies what the problem or opportunity is and presents a concept for a product as a solution.

The output of the product definition stage is a detailed specification of the design and the environment it will operate in. The process specifies the algorithms and functions, the external and internal interfaces, the block partitioning, and the data flows. The product definition consists of waves of refinement, where the first wave begins by defining the specific implementation characteristics, some in great detail and others in less. Each successive wave further defines the details of the product, working toward the final project.

If the product definition is detailed enough, the implementation stage focuses on coding the design and meeting the physical and timing characteristics. Quite often system-level design and architecture changes are made during the implementation stage. Development teams use many different mechanisms for developing the final definition of the design that the implementation teams will use, but the most common mechanism is the system model. A system model is a high-level representation of the design that can be simulated to explore possible architectures or functional trade-offs.

In the early stages of system-level design, high-level models of processors, buses, or interfaces help identify performance bottlenecks. The focus of the system model at this point is interoperability and a fast turnaround in modifying and simulating various scenarios. System models are often thrown away after architectural exploration or functional experiments are completed. This is the first occurrence of fragmentation in the development process. While it is usually not feasible to develop a complete FVP-like model for use in early architectural analysis and system design, developing a system model in a way that enables an FVP to evolve is highly beneficial.

This chapter focuses on the idea generation and product definition stages of the development process and the role verification plays during these stages.

An FVP versus a System Model

System models have been used for many generations of ICs. What is different about these system models and an FVP? In one word: verification. System models are developed by architects and system designers who often view functional verification as an afterthought, if it is considered at all. FVPs are developed by verification teams in coordination with system designers, with the primary goal of unifying functional verification.

The system models of the past have been ineffective for verification for a number of reasons, partitioning being one of them. Architects and system designers are not concerned with following a design partitioning that matches the intended implementation. Algorithms and architectural elements are easier to model as they exist functionally, but these functions may be partitioned in a different manner than the final implementation. If the partitioning of the model does not match the partitioning of the implementation, it is difficult to reuse the model for verification uses, such as reference models or integration vehicles.

The format and level of abstraction at which the model was created also makes them ineffective. System models are written at very high behavioral levels or at very low cycle-accurate levels. Behavioral models are great for speeding verification, but if they are at too high a level, they lack the implementation details needed for verification. If the models are written at too low a level of abstraction, the verification speed is not improved and it is difficult to keep these models in sync with design changes. The preferred level for verification is the transaction level, which provides the right mix of speed and detail.

Another reason system models have not been used in verification is completeness. Architects and system designers are usually concerned with only a portion of the total design. Design parts, such as standard interfaces or service blocks, are not a major consideration for architects, so they are not modeled or modeled inaccurately. Verification development requires a nearly complete model as an executable specification. A system model is also not kept up-to-date. Architects and system designers only focus on the very early stages of the project, and who maintains the model in the later stages often becomes an issue.

An FVP differs in that it is modeled from the beginning with a focus on verification. This does not mean that architectural analysis and system design are not considered or supported, but rather an FVP supports architectural analysis, system design, and software development in a framework that unifies these tasks with the verification process. The FVP is partitioned along the same borders as the main implementation blocks and is developed at the transaction level of abstraction. The FVP is owned by the verification team, which ensures that it is kept up-to-date with design and verification changes throughout the verification process. The FVP is a complete model of the system and testbench.

Costs and Benefits of an FVP

An FVP can greatly aid the verification of many designs today, but for some designs the cost of developing the model could be greater than the benefits received. Teams must have accurate information to determine the costs and benefits before beginning development.

The costs of developing an FVP are measured by the time spent and resources required. The time needed depends on the experience of the developers and the complexity of the design. The larger and more complex the design, the longer it takes to develop a model. Less experienced engineers need more time to learn a new modeling language and the concepts of high-level modeling. Once an engineer has learned the basics, the modeling process can take between 10 to 20 percent of the time required to implement the design. This time can be lessened by reusing models from previous designs or purchasing models already completed.

How long it takes to develop an FVP depends on when it is begun and how resources are used. If resources are taken from other tasks or tasks are put on hold until the FVP is completed, the project will be delayed. If people are engaged earlier in the process when they are usually idle, and the model is completed in parallel with other tasks, less time cost is needed. In addition, if parts or all of the FVP can be combined with other efforts, such as architectural modeling or software development, the cost is reduced. Teams need to decide who will be developing the FVP, when development will begin, and how much can be reused or leveraged from other efforts to calculate the true cost.

The benefits of using an FVP are measured by the time saved and the quality of verification. The up-front investment in time and resources for developing an FVP is offset by the savings in time and resources throughout the development process. The most direct time and resource savings occur when the FVP is used within the verification testbench. Having an early model of the system lets the verification team to do more work in parallel with the implementation team, thereby shortening the overall development time. One of the most time-consuming tasks in developing a testbench is determining the expected results to compare against the results observed from the design. Many verification teams end up embedding large parts of behavioral representations of the design within the testbench or the tests to help determine what the correct expected result should be. Using the FVP in place of this embedded behavioral information saves time and reduces the chance of errors within the testbench.

Time savings can be realized in other ways. Finding architectural bugs earlier in the process reduces the time wasted in redesign and reverification. This time savings is difficult to quantify, since teams do not plan on having architectural bugs. Past experience is the best way to determine what this savings might be. Software teams can also begin development earlier. This may not be a verification time savings, but it might save time on the total project, since software development is often the long path in project development. Time savings can also be realized when the FVP is used as an early-access prototype for other development teams. Providing an accurate executable FVP to a team that is designing this part into their product can save time and respins later in the process.

The decision on whether to develop an FVP comes down to a trade-off between the time invested up front and the time saved during implementation. Teams should ask themselves what is the intended use of the FVP. Will it be part of the testbench environment? Will it be used for architectural analysis, software development, or prototyping? Once the intended uses are known, a team can assess the development costs. It is important to remember that the development costs are not constant. The first projects may come at a greater cost, but as the team becomes more experienced and a larger library of models is developed, the initial costs can be amortized across multiple projects. It is also important to note that once you create an FVP, it may open doors to new ideas.

USING AN FVP

The FVP consists of testbench components, design modules, and interface monitors. The inclusion of the testbench is important, since the FVP provides a design representation as well as the infrastructure and test environment to use the FVP. Testbench components include stimulus generators for driving data into the design, response generators for responding to requests from the design, and application checkers to verify that the FVP is operating correctly. Design modules are the functional models that make up the design. They are developed at the transaction level and follow the refinement and partitioning of the implementation as it develops. Interface monitors verify correct operation by passively monitoring the transfer of information between design blocks.

The FVP is created early in the development process at the same time as the implementation architecture is being defined. The system-level verification team develops and maintains the FVP, which is refined as the design progresses. The first step in creating an FVP is determining its intended use. The FVP is not intended to be a one-size-fits-all solution: Each design and each development process are unique and have their own set of challenges. While the accuracy of the FVP must always be correct, the fidelity of the design may vary. An FVP's detail and fidelity are determined by its intended use. There are three main uses of an FVP:

- As an executable model for software development
- As a reference model for subsystem development and integration
- As a developed model for a design-in team

Each of these uses has different requirements for how the FVP should be developed. A development team must determine which of these uses are relevant to their design and what priority should be placed on each.

Software Development

Developing and debugging software using an executable model can save a project time. Unfortunately, many software teams today have limited resources and are not available to begin work on a new project until the hardware has been developed. In this situation, the benefits of the FVP as an executable model are limited. However, if resources are available, knowing the amount of software to be developed as well as the intended application gives the developer a good understanding of the fidelity required of the FVP and the benefits it provides.

There are three basic types of software applications:

- Service processor
- User interface
- Real-time

Service processor applications are designs where software controls a service processor to handle basic startup and maintenance functions. They require the least amount of fidelity in the FVP. In these designs, the software needs read and write access to registers and memory space to perform such operations as configuring mode registers, error handling, and collecting run-time statistics. The FVP must closely model the software-accessible registers within the system and provide basic register and memory functions. The algorithms, data paths, and processes within the blocks can be a very high-level implementation and non-specific.

With user interface applications, software controls the processor to let a user control and monitor the operation of the system. These designs require greater fidelity than a service processor application. The FVP needs to closely model the software-accessible registers within the system and be able to monitor run-time events as they occur. The algorithms, data paths, and processes within the blocks have to provide the visibility and control required by the software UI.

Real-time software applications are designs where software is directly involved in the functional processes and algorithms of the system. These designs require the highest fidelity. In real-time software applications, the software is tightly coupled with the hardware. The FVP must closely model the software-accessible registers and memory as well as the algorithms, data paths, and processes within the blocks. This modeling has often already been done in the architecture stage of the design.

Subsystem Development

When creating the FVP, it is important to understand the specific needs of individual blocks. Individual blocks either already exist in an implementation form, such as third-party IP or reused cores, or they entail new development. An existing design might not have a transaction-level model (TLM) built for it, so the FVP team must create a TLM, which is used for integration purposes only. A TLM for an existing design should concentrate on fidelity at the interface level and only abstractly model the data path and control processes.

FVP blocks that need to be developed might require more detailed modeling. If the block will use the TLM in a top-down manner to verify the individual sub-blocks as they are designed, the development team should make sure that the TLM partitions the functions as they will be partitioned in the design. If the development team is only using the TLM from the FVP for full subsystem verification, internal partitioning is not necessary.

Design Chain Use

Most systems today are designed by a group of design chain partners. The design for which the FVP is created will most likely be integrated into a larger system by a different team or even a customer. The FVP provides these design-in teams a model to begin developing their systems before the implementation is complete. In some cases, these design chain partners use the FVP as part of their larger system model to test functionality and performance. The FVP also provides an excellent vehicle for demonstrating to customers and third parties that the design is progressing as expected.

FVP developers must understand the needs of their design chain partners. The FVP might require sufficient fidelity for the design-in team to develop their models as well as system software.

CREATING AN FVP

The FVP is created in a top-down manner, following the partitioning of the system architecture. The first step is to identify the modules and the correct hierarchical partitioning. Next, a hollow shell of each module should be created, with a name that matches the implementation module name. External ports for each module are then defined and added to the individual modules. Consistent naming between the model and the implementation is important to facilitate future integration.

After the modules have been defined, the developer defines the channels to which the modules are interfaced. Again, careful consideration and planning are necessary to provide common interfaces that can be used throughout the process. Once the modules and channels have been defined, each module has its individual processes defined. The modules are defined to be functionally correct representations of the corresponding implementation. Separate threads for parallel processes are used with the state and stored in member variables.

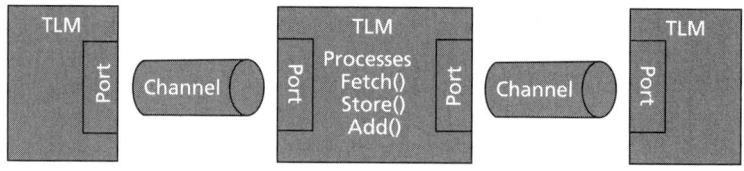

Figure 20.   Creating an FVP

Creating a Transaction-Level Model for an FVP

You can create the TLMs of the FVP from scratch, from other TLMs, or from other model formats. You can create them from behavioral models, analog models, or algorithmic models.

Creating a TLM from a Behavioral Model

Often a subsystem has already been implemented when FVP creation begins. In this case, a TLM may not have been created for the subsystem, or the only model available is a behavioral model. A TLM can be created from a behavioral model by surrounding the behavioral or implementation model with a transaction-level wrapper. This wrapper translates the behavioral interface into a transaction-level interface that can be used in the FVP. The transaction-level wrapper captures the output of the model and calls the SystemC transaction functions across the channels. The transaction-level wrapper also receives transactions across channels from other subsystems and converts the transaction information into a form the behavioral model can utilize.

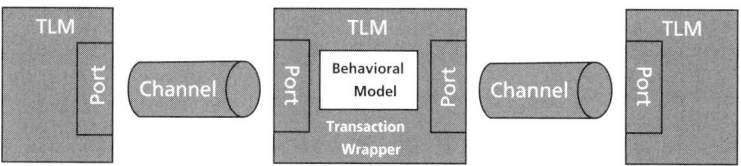

Figure 21. Creating an FVP from a Behavioral Model

Creating an Analog FVP TLM

Analog designs do not operate at the traditionally defined transaction level. Analog designs operate in a continuous-time manner. A transaction in the continuous-time domain can be thought of as a sequenced response over the time unit. Thus, a TLM of an analog component looks very similar to a standard continuous-time model. Analog algorithms are often first represented as high-level C functions and are refined down to behavioral models that represent the implementation of the design. These models accurately reflect the continuous-time nature of an analog circuit.

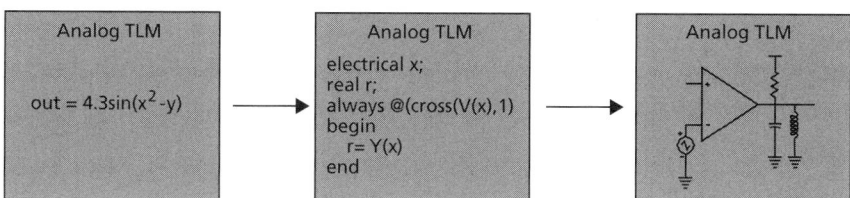

Figure 22. Creating an Analog FVP

The important factor in creating analog models for an FVP is accurately modeling how they will interface to the other subsystem models. In an analog model, the interfacing between analog blocks is done in the continuous domain.

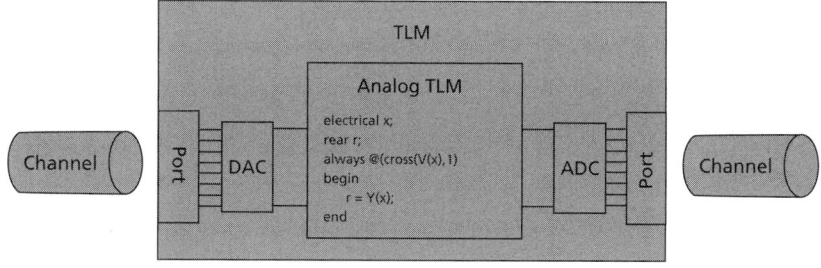

Figure 23. Creating an FVP from an Analog Model

The interfacing of digital and analog subsystems is usually accomplished through digital to analog (D-to-A) or analog to digital (A-to-D) converters. The D-to-A or A-to-D converter transforms the continuous-time algorithm into a signal-level digital interface. This digital interface can then be translated to a transaction interface to enable connecting to digital TLMs.

Creating an Algorithmic Digital TLM

Algorithmic digital subsystems are similar to analog subsystems in that they are first modeled as continuous algorithms. Algorithmic digital subsystems are refined first to fixed-point models and then to discrete-digital models. These subsystems interface between several different domains. Often they interface to the analog subsystem, where the interface is modeled in a continuous-time nature to start. Then, once a D-to-A or A-to-D converter is placed at the interface, the algorithmic subsystem is modeled as a fixed-point representation.

Algorithmic digital subsystems interface to the control-digital subsystems in the FVP. This interface defines how the subsystem is configured for operation and provides the user with statistical information. It is usually a standard bus interface and is easily modeled as a TLM.

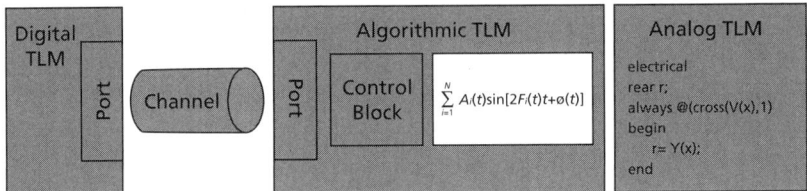

Figure 24. Creating an FVP from an Algorithmic Model

Creating Stimulus and Response Generators

Once the FVP TLMs are created, the model must be verified. Stimulus generators drive data into the model, and response generators provide accurate responses from the model. The generators directly interface to the FVP at the transaction level. Separate drivers for controlling handshaking and signal timing are not required. Stimulus generators can consist of directed tests, random stimulus, or test vectors. Response generators provide the FVP models with application-accurate responses to requests. They can consist of TLMs from external components or be developed similar to the FVP models.

Creating Interface Monitors

Interface monitors are placed between the TLMs, inside the FVP. Interface monitors check the signaling on interfaces between blocks but, since there is no signaling at the transaction level in an FVP, the interface monitors observe the transaction interfaces, check higher level protocols, and collect transaction information. The information collected is useful for debugging the FVP, because interface monitors can identify where a transaction was incorrectly received, narrowing down the source of a bug. The transaction information is also useful for measuring interface coverage within the FVP.

Creating Architectural Checkers

Architectural checkers verify that the FVP meets the specified architectural, functional, and performance requirements. Architectural checks can consist of performance monitors, comparisons to behavioral models, or comparisons to algorithms or data path models developed in signal-processing workbenches.

VERIFYING THE FVP

One of the most complicated tasks in developing the FVP is verifying that it operates correctly. Because the FVP is utilized as a golden reference model, it is imperative that it is functionally correct or the implementation will be flawed.

There are three layers that need to be verified to determine the accuracy of the FVP:

- Performance and functional requirements
- Behavior
- Algorithms and processes

The methods for verifying the FVP are similar to verifying the implementation.

Stimulus Generation

Stimulus generation is a mix of random and directed methods targeted at early architectural and performance verification, along with interface tests to smooth the integration process. The three layers of FVP verification are encompassed in the methods used for stimulus generation.

It is important to verify the performance and functional requirements early in the development process, so architectural errors do not cause redesign later. This testing is done with directed tests to verify architectural behavior in a basic, isolated manner. Early random testing may introduce too many variables, clouding the verification and making debug difficult. Once the directed tests have verified correct basic operation, directed random testing is used to test special cases and stress conditions for the architecture.

The FVP could be meeting the performance and architectural goals, but its behavior does not match the architect's intention. Directed and pure random stimulus, along with behavioral monitors placed on individual modules of the FVP, determine whether a unit is behaving in an improper manner, such as dropping more packets than it should or bypassing stages in a pipeline.

Often the stimulus that was used to verify the algorithms in isolation can also be used in a directed manner to prove the correctness of the algorithms in the FVP. Otherwise, directed tests are used to stimulate the basic operation and known corner cases of the FVP. Directed random tests are then used to cover unknown cases.

Architectural Checks

Architectural checks monitor and capture the response of the FVP to the stimulus. The three layers of FVP verification are encompassed in the methods used for architectural checks.

The verification team works with the architects to define the functional and performance requirements for the system. They then develop checkers to verify this operation. Functional requirements can include calculation accuracy, event ordering, and correct adherence to protocols. Performance requirements can include bandwidth, latency of operations, and computation speeds. These checkers can be self-checking or require post-processing. In any case, they are the basis for the architectural assertions to be defined later.

Architects often model some of the more complex operations of a design at the behavioral level to create the best solution and to measure trade-offs. These models are used as reference models for the FVP to verify that the intended behavior of the design is implemented correctly.

Many functions of a design can be described as algorithms or simple process descriptions. These algorithms and processes can be represented in many different forms or languages. These functions within the FVP can be verified by either embedding the representations into the models with a transaction-level wrapper, or by running the representations in parallel to the model and verifying that the outputs are equivalent.

It is important to consider the scope of the testing when developing architectural checks. The architectural checks are meant to test the operation of the entire design to the intended specification. Implementation details of the subsystems are tested by the individual subsystem teams, and system verification tests the correct interface with real-world stimulus.

Advanced Verification Techniques

Advanced verification techniques, including assertions and coverage analysis, speed the process of verification. Assertions are not commonly thought of as a technique used in developing and verifying a system model. The common perception of assertions is of implementation-level Property Specification Language (PSL) type checkers. Although standard PSLs and formal tools are not available today for high-level models, the technique of embedding checks within the design of a high-level model is still useful. At the FVP level, these assertions can be coded in SystemC using standard print statements to indicate detected errors.

Application coverage monitors are included with application assertions to verify that each of the functional requirements has been stimulated before the FVP is handed off to the next team. Architectural coverage should be measured after the basic system tests have been run, along with some random tests. Interface coverage is also used throughout the development of the FVP. When it is first developed, the FVP is transaction-based, so verifying handshake and signaling coverage is not relevant. However, interface coverage is important for identifying which stimulus has been applied to the design and correlating the responses. Other advanced verification techniques that can be used are discussed in other chapters.

Chapter 8

Control Digital Subsystems
Verifying large digital designs

The second phase of the UVM is subsystem development. In this stage, separate teams design and verify each of the individual subsystems that make up the final system or device. The subsystems can be separated by function or by design domain, allowing parallel teams to focus on development of smaller individual pieces. Today's modern SoC might consist of subsystems from one of three design domains: control digital, algorithmic digital, and analog/radio frequency (RF). Although each of these subsystems are developed and verified in different manners, they are unified in the UVM through the FVP. Each block can use the models from an FVP as a reference model to compare to during verification, and each subsystem can use the FVP as an early integration vehicle. The following three chapters discuss each of these subsystems in more detail.

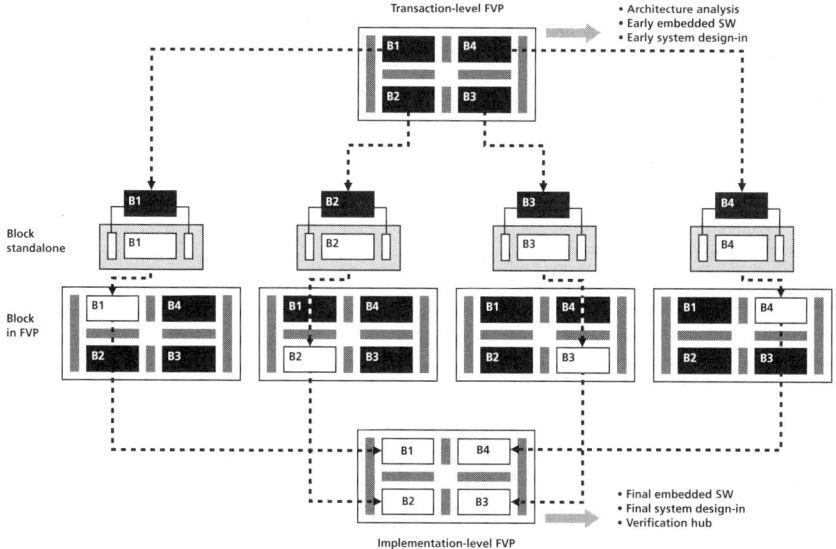

Figure 25.   Subsystem Development in the UVM

Control digital subsystems are the most predominant design domain in ICs today. Many of today's devices consist entirely of one or more control digital subsystems. Control-based digital subsystems are developed from the specification of a control process. The designs may contain data paths but are not based on algorithmic processing. The complexity associated with massive digital subsystems has been the focus of much of the functional verification efforts to date. Specialized tools for adding assertions and measuring coverage, along with specialized verification languages and methodologies, have been developed to address this. Unfortunately, this focus has been the source of much of the fragmentation found in functional verification today. These focused methodologies and tools force the verification team to start development from scratch, isolate the verification of the subsystem from other design domains, and provide for little, if any, reuse during integration and system verification. The UVM removes the fragmentation associated with methodologies focused solely on the digital subsystem level by using the FVP to unify the different stages and design domains. The verification of a control digital subsystem is broken into four phases in the UVM, as described in this chapter.

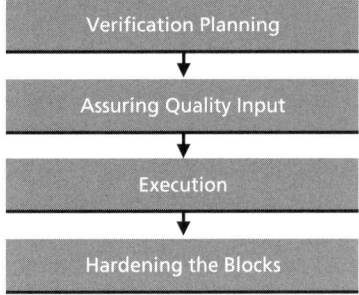

Figure 26.   Four Stages of Control Digital Subsystem Verification

STEP 1: VERIFICATION PLANNING

The first stage of the control digital subsystem verification process is strategy and planning. A level of verification planning was already done during the architectural definition stage when the FVP was created. At that time, planning focused on the system as a whole. Subsystem verification picks up from this system-level plan and refines it down to individual subsystems. The planning process for a control digital subsystem has four steps: setting goals, defining strategy, planning tactics, and defining the measurement and analysis mechanisms.

A team should not wait until they have all the necessary information to begin planning. Instead, advanced verification teams begin planning with the information they have. As new information becomes available, the plans can be refined or updated. The planning process needs to include a feedback system to make sure that the individual goals obtained, from the lowest feature or test level up to the ultimate product goal, tie together.

Figure 27.   Verification Planning Flow

Goals and Objectives

The verification planning process is driven and facilitated by the development of a test or verification plan document. A successful verification plan starts with the end in mind. The plan begins by defining the goals and objectives for the verification of the specific control digital subsystem. The goals articulate what will be achieved, when it will be achieved, and how success in meeting the goals will be measured. Each team will have their own particular goals that matter the most to their organization. Unless the project is intended solely for research purposes, schedules and delivery dates often influence a verification project.

Effort is often overlooked when specifying verification goals. It is important to determine whether the project should expend a lot of effort on creating reusable quality results or whether a "just get the job done" approach is more appropriate.

Completeness is another area overlooked. Is it imperative that the complete subsystem and all its features be verified at this time or are there certain must-have features that are needed while others can be verified later? Every verification plan should specify the quality of verification intended for this project. In a perfect world, every project would verify to the highest level of quality, ensuring no bugs escape the process. In reality, there are cases where lesser levels of quality may be acceptable. For instance, if there are features that will be tested in later stages of the process, such as system integration or prototype stages, complete testing might not be necessary at this time.

Strategy

Once a clear set of goals have been articulated, the next step is to develop a verification strategy. Strategy is an overloaded term these days. For the purposes of this discussion, we separate strategy from tactics. Strategy is the general approach to be taken to meet the specified goals. Tactics are the methods and tools used to implement the strategy.

To develop a clear strategy, the team needs to first review the goals, the design, and the environment the project must operate in. At this time, the team should identify design issues and obstacles and prioritize them. Possible issues or obstacles include the size of the design, the schedule, the number of features to test, and how to determine when you are done. The most creative part of the planning process is developing strategies that address the identified issues and attain the desired goals. Groups generate ideas and strategies in different ways, varying from brainstorming sessions to mechanical problem-solving practices. Regardless of the process, it is important to be open to new ideas and keep the big picture in mind. Examples of strategies for addressing common issues include:

- Partitioning a larger design into smaller segments for isolated testing
- Cutting the schedule by staging deliverables, or purchasing or reusing older environments
- Using random testing or automation to verify as much of the state space as possible
- Using specific coverage metrics to verify that the design is ready to tape out

Tactics

After the verification strategies have been determined, the team can focus on the tactics for addressing the strategy. The development of tactics is where the real "go to battle" planning is done. One can think of it as similar to developing a playbook for a sports team or military unit, detailing the intended use of the people and technology, the environment and infrastructure, and the coordination and communication methods. Having a comprehensive playbook keeps developers in sync and provides guidance to those who join later or work in associated groups.

Listing the specific responsibilities and activities associated with each tactic lets the team know the training and development needs as well as provides clear directions for resources. In a similar fashion, detailing which tools and technologies will be used with each tactic leads to smarter purchasing and utilization decisions.

Perhaps the most important part of developing a tactical plan is detailing the environments and infrastructure to be used, such as a description of the testbenches to be developed as well as the APIs and interfaces to use. A project might have many people moving on and off the project. Detailing the basic testbenches and infrastructure enables them to come up to speed quickly and operate efficiently. Finally, the tactics should describe the processes and methods for maintaining communication throughout the project. Having all this information together helps the entire team understand what needs to be done and how it will be accomplished.

Measurement and Analysis

The final part of the verification planning process is measurement and analysis. This phase details each goal down to the task level and articulates how to determine whether a goal is obtained. Goals can include the development of testbench components, stimulus scenarios to be tested, features or functions to be verified, and integration with other components. The details could specify the components to be built and tested, directed tests or stimulus streams to be created, monitors or assertions to be placed in a design, or specific integration scenarios.

Identifying the tasks and features is only one part of the measurement and analysis process. Setting goals without knowing the mechanism for verifying the goal is dangerous and can lead to holes in the verification strategy. Assertions or monitors can be used to identify that a directed test verified the intended feature. Automated coverage tools can be used to identify that a special coverage scenario was met. There should also be a way to track the status as the project progresses. Correlating regression results or coverage analysis information to the goals not only helps management track project status, it allows the entire team to coordinate and adapt their effort to reach the final goals.

The planning phase is an important first step in any complex process. Although it is the first step, it does not end when the project moves into the execution phase. A team should not begin execution before they have an initial plan, but they should also not wait for the planning process to come to completion before beginning execution. Teams should begin planning with the information available, refine the plan as the project progresses, and never take their eye off of the final goals.

STEP 2: ENSURING QUALITY INPUT

The old axiom of "garbage in leads to garbage out" is very applicable to the verification process. Even the most powerful complex verification environment is of little use if the HDL is so full of bugs that the tests never even get started. Ensuring quality design input from the beginning limits the amount of wasted time and effort for the verification team and allows the project to reach completion faster.

This phase has two main goals. The first is using the right tool for the right purpose. Simple typographic bugs happen in every design and can be found with almost any verification technique, but why expend time and effort on a complex tool to find a simple bug. The second goal is to clear the way for downstream verification processes. Many verification processes have specific requirements for the HDL code or test specific aspects of the design. It is helpful to identify any issues that might impact downstream tools early in the process while the designer is still engaged in developing the code. For example, an incorrect clock domain synchronization issue might require the designer to redesign large portions of the design. Identifying this issue early, before millions of random simulation cycles have been run, alleviates the reverification effort of design changes.

An important factor in this phase is to perform verification while the design is being developed. At this stage, much of the verification effort relies on the HDL designer. After the design is completed, designers move onto implementation tasks, so it often becomes impossible to make the necessary design improvements. Another important consideration is to not get in the designers way during this phase. Verification needs to be automated, easy to use, and have a high payback for a designer. Otherwise, the designer will leave the work for the verification team to complete on their own.

The UVM identifies four specific tasks during this initial phase of verification, described in the following sections.

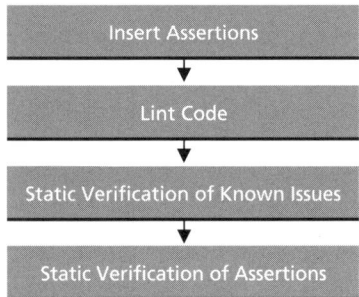

Figure 28.   Ensuring Quality Input Process Flow

Instrumenting with Assertions

Assertions have many uses in the verification process and are discussed in several places in this book. Assertions define an operation of a design and indicate whether the operation is incorrect. Assertions can take many forms, such as HDL display statements, statements written in a dedicated assertion language, checkers that can be instantiated within a design, or pragmas that provide a shorthand indication of comment assertions. Assertions provide a more direct and detailed checking of a design than normal end-to-end checkers.

There are several types of assertions. The system designer develops high-level assertions for verifying architectural requirements in coordination with the FVP. Verification and integration teams develop interface assertions to verify the protocol and operation of external or internal interfaces. Structural assertions verify the operation of low-level implementation structures within a design, such as the correct operation of a FIFO under overflow or underflow conditions or a state machine following the correct transitions during special case interrupts. The designer of a block is usually the only one who can determine what the correct assertion is and where to place it.

The UVM utilizes assertions at each of the different levels of design. Structural assertions play an important role during simulation, integration, and final verification. The designer should place assertions while the design is being developed. Advanced verification teams have found that having another engineer place structural assertions into a design can be very difficult and time consuming, since the engineer is not familiar with the design. Advanced teams have also learned that once a designer completes coding a block, it is difficult to motivate the designer to go back and insert structural assertions. The correct time to place structural assertions is when the designer is writing the code.

Motivating a designer to place structural assertions within the design can be difficult. Designers do not always see the value in placing assertions, since they believe that the design is correct to begin with and verification is not their responsibility. Overcoming these objections can be difficult. Advanced teams have found that showing the designer the benefit of having less time spent doing verification later in the process is one motivation. Another is the ability to more easily identify where in a design a failure has occurred. The designer should spend as little effort and time as possible in placing assertions, because the easier it is to do, the more likely the designer will place the assertions in the correct places. The UVM recommends that structural assertions be placed using libraries of predefined checkers or simple pragmas placed in the language.

Linting

Linting is the most basic form of static verification analysis. There are several different types of linting technologies available today, which are discussed in detail in Chapter 13. Designers should do linting up front in their development process. The most basic forms of linting can identify simple typographic errors or basic implementation bugs, such as unconnected buses. Finding these errors quickly here is more efficient than in the front end of an advanced tool or in the process of debugging a simulation failure. More advanced forms of linting can be used to identify code issues, such as synthesizablity or dead code that may cause tool issues later in the verification process.

Verification engineers or the HDL designer can run linting tools, but usually a designer needs to act on the results. The more basic linting tools should be available for designers to run at any time while they develop their code. The more often the designer runs these tools, the better the quality of code delivered to the verification team will be. Motivating the designer to run these tools requires that the output be of good quality, without many false or incorrect violations. The tool should be fast and allow the designer to identify the issue within their code quickly. Advanced verification teams have found that the investment in time and effort to use linting pays off over the entire project cycle.

Static Verification of Known Issues

Static verification is verifying a design without dynamic simulation. Static verification is usually based on formal verification techniques that use mathematical principles to verify a design. Formal verification is a large topic and discussed in detail in Chapter 13. Formal verification techniques can be very powerful, but can also be very difficult to use and time consuming. Several specific applications, such as verifying clock domain crossings and state machine reachability, can harness the power of verification in an easy-to-use manner. Using these applications early in the design process before simulation testbenches are available can identify bugs that can be fixed while the designer is still engaged.

Clock domain crossings are a particularly important application of static verification. Today's designs often have many different clock domains and the transfer of information between those domains requires tight synchronization. You can attempt to simulate the different possible interactions between the domains, but it is impossible to verify every possible combination of clock skews. Verifying the synchronization logic is often left to visual inspection by experienced engineers. The UVM recommends using static verification tools once the design is completed to immediately identify incorrect operation and get it fixed. This early detection of known design issues can save effort and simulation cycles to reverify a design after a bug is found late in the project cycle.

Static Verification of Assertions

Static verification techniques can also be used to verify the structural assertions that the designer placed in the design. Verifying user-defined assertions can be more labor intensive than focusing on known applications, such as clock domain crossings. Depending on the assertion, the tool might need to make assumptions about the possible stimulus of the design at the inputs or the operation of blocks within the design that are not visible to the tool. These assumptions can lead to the tool identifying issues that are incorrect because they are based on incorrect assumptions. Qualifying the inputs of a design and localizing the design so that the tools can produce quality results is a very labor-intensive process. The UVM recommends using static verification techniques to verify structural assertions within the design, but large amounts of time should not be invested in the process if the results are not of high value. It is recommended that you focus on violations that are not dependent on input assumptions or black-box logic.

There are three main purposes of static verification. One is to identify bugs within the design from assertion violations. These bugs are found sooner in the process, thereby reducing the amount of resimulation required after the fix is done. The second purpose is to identify incorrect assertions. Quite often an assertion is written incorrectly by mistake. Finding these mistakes early saves debugging incorrect violations later. Finally, it is to identify assertions that are proven to never be violated. These assertions can then be removed if speed is an issue, or tests might not need to be run because the assertion has verified certain logic.

STEP 3: EXECUTION

The third and most time-consuming step in the verification of control digital subsystems is executing the verification plan. During this stage testbenches are developed, tests are written, and the design is simulated in a manner defined in the test plan. A verification team can take two possible paths of execution. The traditional path is to break the design into parts, start verifying the individual parts in isolation, and then integrate them together in a flow that moves from the lowest level, or bottom up, to the top subsystem level. The majority of verification teams today use bottom-up flows. A newer and more advanced approach is to start by verifying the highest subsystem levels first and then break down the design into smaller parts for isolated verification as needed. This top-down approach allows teams to maximize their efficiency through reusing testbenches and models. The UVM supports both the top-down and bottom-up verification approaches.

There are many similarities between the top-down, FVP-based flow and the bottom-up, specification-based flow. Even though the bottom-up flow develops the testbench from the lowest unit level up to the subsystem level, the planning and architecture for the flow is top-down. The only way to create testbenches that can be reused as the design develops from the unit to the block to the subsystem level is to plan ahead and know what will be required at higher subsystem levels. Similarly, even though the top-down flow develops the testbench from the top SoC level down to the lowest unit level, testing and integration are still performed in a bottom-up manner. Both flows use transaction-based testbenches for performance, assertions for easy debugging, coverage for efficient test development, and hardware acceleration for increased performance.

The two flows differ in the development of the verification environment and tests. In the top-down flow, the environment is developed from the highest levels (SoC FVP) down to lower levels, using common models and testbench components. The SoC-level environment is developed first with the SoC-level FVP. The FVP is refined down to the block and sub-block level, and the testbench is developed in the same manner. This enables the verification engineers to use the model to test their code, allows reuse of the models in the FVP at different levels, and promotes parallelism and accelerated integration of the implementation when it is made available.

The bottom-up flow develops the testbench from the lowest level up. The flow reuses testbench components from the lower levels as the design is integrated and tested. Since a common model is not used, reference checkers might need to be developed at each level or linked together. Also, the lack of an accurate model means the tests and testbench components are developed in isolation until the implementation is available. This might result in debugging the implementation and testbench simultaneously.

Testbench Development

The method used for developing a testbench has a dramatic impact on the overall performance and efficiency of the entire verification process. Reuse speeds the development of testbench components. Limiting the amount of application-specific information encapsulated in testbench components facilitates reuse. Raising the level of abstraction makes test writing and debugging much more efficient. Using standard, defined interfaces facilitates the communication between components, making it easier to work at higher levels of abstraction.

High-performance reusable testbenches are based on standard components with a common interface for communication at different levels of abstraction. The basic components are shown in Figure 29.

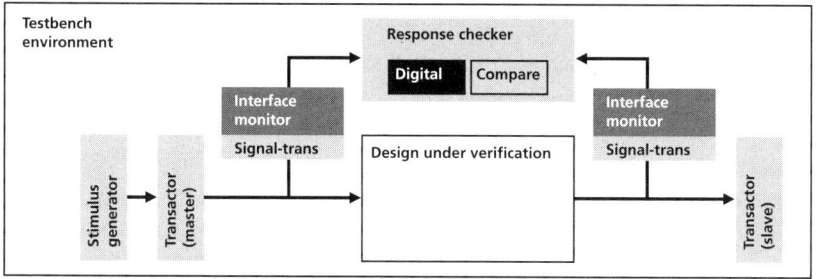

Figure 29.   High-Performance Reusable Testbench

Advanced Verification Techniques

While testbench development and simulation are the most used techniques for control digital subsystems, other advanced verification techniques also play an important role.

Assertions in Simulation

Assertions are used extensively in coordination with simulation and the testbench. Architectural assertions developed in the FVP that are associated with the subsystem are reused if possible. In many cases, the assertions are no longer complete, since the design has been partitioned. These assertions are then disabled at lower testbench levels. Interface assertions are also reused from the FVP for external subsystem interfaces. Interface assertions are added to internal interfaces as they are defined during micro-architecture. These assertions are usually part of the interface monitor testbench components and are specified outside the implementation code.

Coverage

Coverage techniques are used throughout the control digital subsystem development. Application-level coverage defined at the FVP is usually measured at the SoC model and integration stages, since it often encompasses the entire SoC. New application-level goals for just the control digital subsystem are defined during micro-architecture of the subsystem and are implemented in testbench code as monitors. These subsystem-level application-coverage goals are measured at the end of subsystem testing. Interface coverage is first defined in the system test plan as part of the definition of the transaction tax-onomy. The verification team defines the transaction types used at the subsystem level and also defines goals for the types and sequences of transac-tions driven into the DUV. Specific correlation goals between stimulus and response are also defined at this time. Interface coverage is first measured when block-level testing is completed. The team verifies that all specified types of transactions have been stimulated, along with a large percentage, if not all, of the combinations of transaction types. Specified correlation goals should also all be measured before block verification is considered complete. Interface coverage is also measured at the subsystem level to verify that all transaction types have been simulated, along with a subset of the possible transaction sequences.

The designer first adds the structural coverage monitors along with the structural assertions. In many cases, assertions can be used to monitor correct behavior as well as incorrect behavior. The verification team adds to these coverage monitors in the form of structural assertions when they receive the implementation from the design team. Structural coverage is measured when block-level testing is completed. Block-level testing focuses on the specific implementation features of the design; this is where structural coverage pro-vides the most information. Structural coverage information is collected after all tests have been run and passed, since it slows down the run times and does not provide accurate information until the tests pass. Code coverage is also run after block verification is complete. The verification team identifies holes in the structural and code coverage and investigates to determine whether the stimulus is lacking, the design is in error, or there is dead code. Coverage is an iterative process in which the results are analyzed and modifications are made to the tests or implementation until the team has addressed all coverage holes.

Acceleration on Demand

Acceleration is used in the control digital subsystem to run long tests faster and run efficient test regressions. Many test sequences take long peri-ods of time to set up because of deep memory queues and complex control space. In these tests, acceleration reaches the desired states faster than stan-dard simulation techniques. The testbench can be run on the simulator in lockstep with the accelerator or, for faster performance, the testbench can be compiled into the accelerator. Testbench components should be developed with acceleration in mind.

After a few tests have been run on the implementation and are passing, an automated regression environment should be established. A periodic regres-sion is run to ensure that changes to the implementation or the testbench environment have not broken existing tests. As the number of tests grows in the regression, acceleration and server farms complete the regression in a timely manner. Server farms are used to run small run-time jobs in parallel with automated scripts. Accelerators are used to run long regression tests and run groups of shorter tests quickly in a serial manner, requiring a smaller server farm and fewer licenses.

Top-Down FVP-Based Flow

The top-down FVP-based verification methodology for a control digital subsystem is broken into three parallel tracks that converge throughout the process. The first track is the further development of the FVP. The modeling team updates the FVP as the implementation is refined. The FVP is used by the other subsystems and possibly at other levels of testbench hierarchy. The modeling team uses the FVP to develop models of other subsystems and reuse TLMs for reference models.

The second track is the implementation of the design. The design team uses an HDL to implement the design, starting with the development of small implementation units. The development team adds structural assertions and verifies these small units to a basic level of operation. The team then integrates these units into design blocks and provides them to the test team for verification.

The third track is developing stimulus. The test team begins by writing a test plan focused at the subsystem level. The test plan is a continuation of the system-level test plan, refining the test strategy and the plans for transactions, assertions, and coverage. The test team executes the test plan on the blocks provided by the design team to verify their functionality. The design team continues integrating units into blocks for the test team to verify.

The UVM speeds the development and verification of the control-based digital unit in several ways. Reusing models and testbench components from the FVP and other subsystems decreases the development time. Test development at the transaction level and the use of the FVP as a testbench raises the level of abstraction and speeds the test time. Adding assertions and reusing them throughout the verification process speeds the debug time and increases the quality of the design. Adding hardware acceleration into the verification process increases the overall performance of the verification process.

Figure 30 shows a top-down UVM-based flow for two blocks developed independently and integrated into a subsystem.

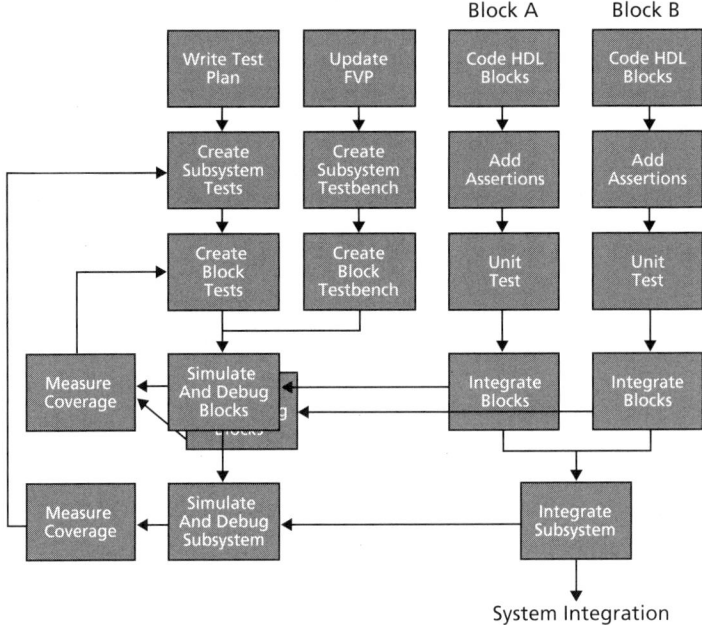

Figure 30.   Top-Down UVM Flow for Control Digital Subsystems

Bottom-Up Specification-Based Flow

The bottom-up specification-based verification methodology is broken into two separate parallel tracks that converge throughout the process. Separate verification and design teams each begin work from the lower blocks of the design and work their way up integrating and testing. If the development teams do not have separate design and verification teams, the two tracks can be performed in serial developing the code first, then developing the tests and verifying the design. The separate serial track is obviously slower and less efficient, but requires fewer resources.

The first track is implementing the design. The design team uses an HDL to implement the design starting with developing small implementation units. The development team adds structural assertions and verifies these small units to a basic level of operation. The team then integrates these units into design blocks and provides them to the test team for verification. Once all the design blocks are developed, they are integrated into the subsystem for integration testing.

The second track is developing the tests and testbench environment. The verification team begins by developing a detailed test plan for each block in the subsystem and for the subsystem as a whole. The team then creates the block-level testbenches in a manner that allows for reuse at the subsystem level. Once the block-level testbenches are complete, the verification team writes block-level tests and waits for the implementation blocks to be delivered by the design team. When the blocks are delivered, they are verified individually. The team measures coverage and adds tests as required to meet the specified goals.

After the blocks have been verified individually, the verification team creates testbenches to integrate and test the blocks together. Depending on the size and number of blocks, there might be many integration steps or just one integration into a complete subsystem. The testbenches are created with parts from the block-level testbenches where possible. The verification team develops integration-level tests and verifies the integrated blocks together in the testbench. Once the integration testing is complete, the subsystem is tested inside the FVP provided by the SoC team to verify that it works with other subsystems.

Figure 31 shows a bottom-up UVM-based flow for two blocks developed independently and integrated into a subsystem.

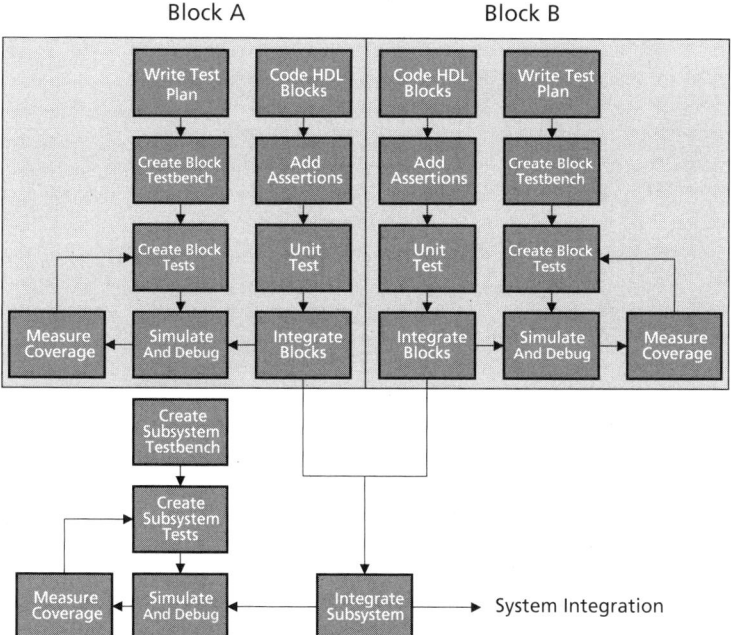

Figure 31. Bottom-Up UVM Flow for Control Digital Subsystems

STEP 4: HARDENING THE BLOCKS

After interface and feature testing is complete, and constrained random testing has run for multiple hours of simulation, the block is ready to be integrated into the subsystem. In addition to integration testing, the block is hardened by verifying it inside the FVP and verifying that it is ready for use in acceleration, emulation, or prototype. If a top-down FVP-based flow is being used, the block is verified against the FVP by running the FVP test suite against the prototype with the implementation block replacing the model. Transactors are added to the block to translate the transaction layer down to the signal layer, as shown in Figure 32. Adding the implementation block can slow down the simulations and require limiting the number of tests run. The tests stress features that cross blocks, such as pipelines and multiple encapsulations.

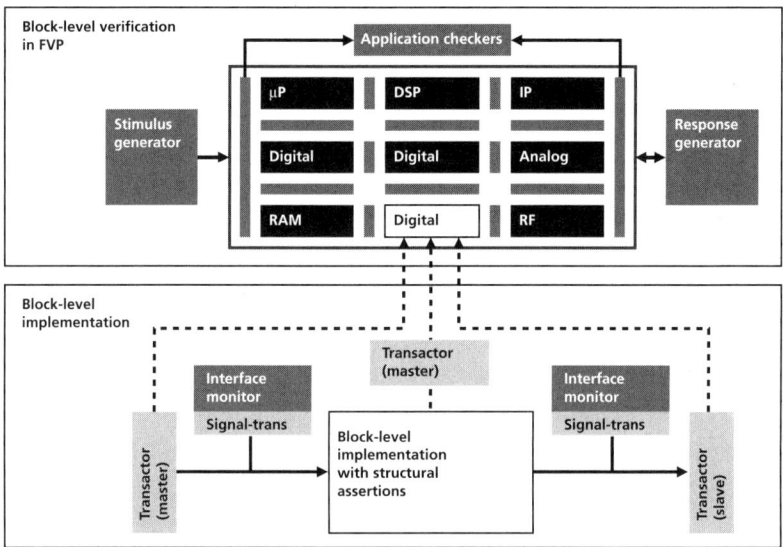

Figure 32.   Integrating a Subsystem into the FVP

Large design blocks need the performance of a hardware accelerator to run long test sequences or to test multiple blocks integrated together. Once the block is stable, it is hardened by running it in isolation through the mapping process on the accelerator. This allows for the early detection of mapping or library issues that could stall the later use of acceleration. In a similar manner, the design is run through any synthesis or mapping processes needed for FPGAs or hardware prototypes used in system verification.

Chapter 9

Algorithmic Digital Subsystems
Verifying algorithms

Algorithmic digital subsystems are found in designs such as digital signal processors (DSP), wireless communications devices, and general data path subsystems. The development and verification of these subsystems has traditionally been left to the few specialists who understand the complex algorithms and know how they should be implemented. Today's SoCs combine these algorithmic subsystems with standard processors and interface blocks to provide a complete solution for the customer. The combination of these subsystems requires more than just a few specialists to understand and perform design and verification.

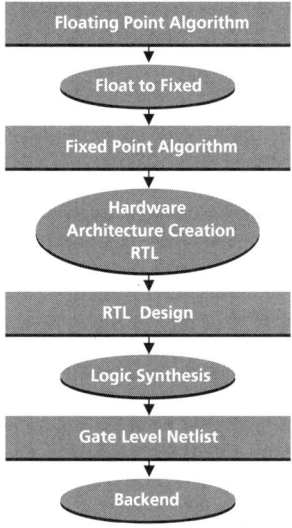

Figure 33.   Development Process for an Algorithmic Subsystem

Algorithmic digital subsystems are developed in a top-down process that refines the algorithm to a specific implementation, as shown in Figure 33. The UVM speeds up the verification process by closely matching the refinement process along with the FVP. The verification of a control digital subsystem is broken into four phases in the UVM, as described in this chapter.

STEP 1: VERIFYING THE ALGORITHM

Algorithms can be developed in many different ways. Some originate from a known industry standard, such as image or sound encryption and decryption algorithms. The designer might modify these algorithms for easier implementation for different levels of accuracy, but the main function remains the same. Verifying these types of algorithms is straightforward. Stimulus is provided to the inputs of the algorithm and the responses are compared with output from other existing algorithms. The main purpose in this verification is ensuring that the algorithm is compliant with the standard.

Algorithms can also originate from combining and modifying existing algorithms. The processing of data received over a communications channel often must undergo multiple transformations before it can be utilized in an SoC. These transformations might be implemented in individual components that need to be combined and integrated into an SoC device. The communications channel may support new characteristics, such as higher speeds or less data loss, which requires modifying an existing algorithm. Verifying these algorithms requires verifying the integration and interaction of the algorithm with the entire subsystem. Testbenches are developed from algorithmic models of the entire system. The system representation is verified by simulating real stimulus, such as a communications channel, and then measuring the output for accuracy. The main purpose of this verification is to ensure that the algorithm provides the required system function.

While it is rare, some algorithms are developed from scratch. Not every algorithm that is developed is destined to become an industry standard, but they all must originate from initial development. Algorithmic development from scratch is similar to the architecting of any complex system in that it is driven by the designer's thought process. In whichever way the new algorithm is developed, it needs to be verified both in isolation and in the context of a system. Verifying the algorithm in isolation ensures that the algorithm meets its desired function over the entire range of operating conditions. Verification within the system ensures that the algorithm provides the intended functionality in the context of the entire system. The main purpose of this is to provide a quality algorithm that can be used in many different systems.

Algorithm Development in an FVP

An FVP can provide the ideal environment for developing and verifying an algorithm. The FVP provides accurate models of the entire system, along with stimulus generation and response checking. A modern mixed-signal SoC provides functionality in both the continuous time domain implemented with analog circuitry and the discrete time domain implemented in digital circuitry.

The partitioning between the two domains is often blurry at the early stages of system design and architecture. An algorithmic subsystem most often exists in this blurry partitioning range.

The FVP allows the architect to refine the system architecture to make the correct partitioning between the continuous and discrete domains. An FVP supports the development and verification of both domains together in one system representation. The algorithmic subsystem can be modeled to operate in either domain until final decisions are made. Converters between the domains can be used to isolate the analog and digital circuitry as the architect sees fit. Once the architecture is completed, the FVP should contain a mix of models that interface at the transaction-level, along with algorithms and converters that interface to the continuous time models.

Algorithmic Models

Algorithmic-based subsystems are most commonly used today in communications and multimedia systems. The system and environmental effects on these types of designs are not as easily predicted as they are in control-based designs. This unpredictability increases the likelihood of an error in the algorithm not being detected until system integration testing. Algorithmic development teams need to verify the intent of the design before implementation is performed. There are too many variables to wait for an accurate implementation before beginning verification.

Algorithmic-based subsystems are developed by either modifying some blocks in an existing system to provide a new function or reconfiguring the blocks of existing systems to provide a derivative function. Both these processes require accurate modeling of the individual blocks at different levels of abstraction. A broad range of building blocks is necessary for developing and verifying the subsystem. Communications and multimedia applications are usually based on standards and protocol layers. A broad up-to-date library of standard algorithmic system components is required for accurate development and verification.

The algorithmic subsystem is modeled in the FVP based on the application. The FVP is a transaction-level model of the system, but algorithmic subsystems often operate and interface in a more continuous-time domain. To correctly model an algorithmic subsystem for the FVP, each interface should be modeled in the most efficient manner for passing information. Algorithmic subsystems interface to other mixed-signal subsystems where a continuous time-based interface is most efficient. The simulation of these mixed-signal interfaces is discussed in Chapter 8. Algorithmic subsystems also often interface to control-based subsystems. These interfaces are defined at the transaction level similar to control-based digital subsystems.

STEP 2:TESTBENCH DEVELOPMENT

As the algorithmic subsystem is refined from the algorithm to the fixed-point representation to the final RTL or gate-level representation, the accuracy of the algorithm is affected. The testbench environment must verify that at each stage of development the accuracy of the algorithm is still acceptable. Verifying this accuracy causes the testbench to become a mix of models at different abstraction levels. The environment might continue to be modeled at the same level as the block under development moves toward the implementation level. Unlike a control digital subsystem where the responses of the subsystem are verified against an expected result, algorithmic subsystems require analysis of the responses to verify that the response is still within specification. This is accomplished with detailed instrumentation that captures the response of the subsystem and produces diagrams and calculated error rates for the developer to verify. This instrumentation is at the heart of the testbench environment.

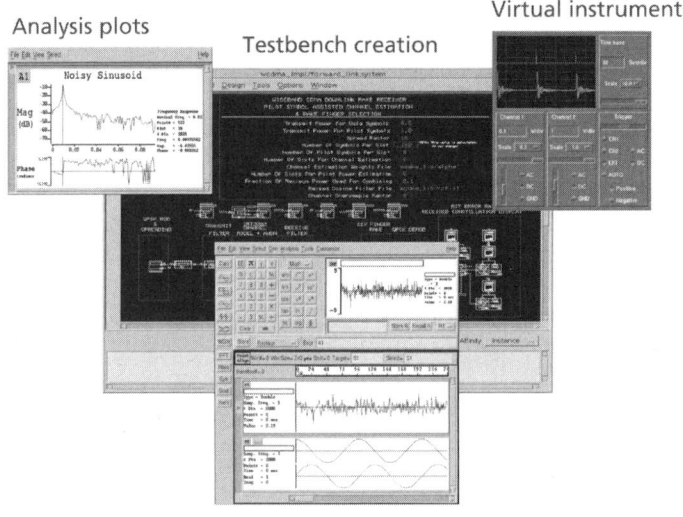

Figure 34. Algorithmic Workbench

Algorithmic digital subsystems verification is most commonly started in a separate but common integrated workbench environment, like the one shown in Figure 34. These environments provide easy intuitive user interfaces for selecting and connecting the various components that make up the testbench environment. The workbench environments should include large libraries of standard components that can be used as a template for new development or as models for simulating the surrounding environment.

The algorithmic development workbench also includes tools for stimulus generation, data collection, and result analysis. The generators provide the developer with control over standard stimulus streams found in communications or imaging environments. Data collection tools can sample and store data at various rates and depths depending on the developer's needs. The analysis tools provide calculators and graphing capabilities to quickly identify results over a wide range of operating conditions.

As the subsystem development progresses, integration becomes more important. The testbench continues to use signal stimulus generators and instrumentation, but also begins to interface with the other subsystems. The design is updated in the FVP so that integration testing can be performed before the implementation is final. The software team develops code for the application on an ISS of the processor core. The analog/RF subsystem is integrated to detail the behavioral effects of the two subsystems. Finally, the control path for the subsystem is verified by attaching the implementation to the FVP in place of the TLM. Verifying the interaction with each subsystem in the FVP before integration speeds up the final SoC integration.

STEP 3: VERIFYING THE FINAL IMPLEMENTATION

Once the algorithm has been developed and verified and then converted to a fixed-point representation, it is ready to be converted to a standard hardware implementation format. Automated synthesis tools can automatically convert the fixed-point algorithm to hardware-gate netlist. These automated synthesis tools provide limited control over many of the important characteristics of the resulting implementation. Developers might choose to not use these tools if they require tight control over the accuracy of the final implementation results. Developers also find that the results of these synthesis tools are difficult to debug due to the readability of the machine-generated code. If bugs are encountered in the implementation netlist, it is near impossible to locate the cause in the generated code

Most developers choose to hand-code the hardware implementation of the algorithm at RTL from the fixed-point algorithm. This provides the developer with tight control over important implementation characteristics and provides code that can be debugged more easily. No matter how the implementation is developed, the chance of a bug entering the process is high. Verification of the final implementation focuses on ensuring that the hardware implementation of the algorithm accurately reflects the intended algorithm. The first step in this process is placing the hardware implementation back into the algorithm workbench used to verify the original algorithm. Care must be taken when implementing the RTL so that it interfaces correctly in the workbench environment. Unfortunately, the RTL model simulates at much slower speeds than the algorithmic equivalent. This may result in reducing the amount of testing performed in the environment to just the most critical.

The hardware implementation of the algorithm should also be verified in a standard HDL-simulation environment. The algorithmic subsystem should be simulated with the connecting digital subsystems to verify connectivity and interoperability. The FVP or a modified existing control-digital subsystem testbench can be used for this testing.

Advanced Verification Techniques

The initial development of an algorithmic subsystem concentrates on the algorithmic and fixed-point representations of the design. Advanced techniques, such as assertions, coverage, and acceleration, are not used during these early stages. The conversion of the fixed-point representation to an implementation can be done with automated synthesis tools or can it be done in a more traditional hand-coded method. If the design is synthesized or hand-coded, interface assertions are added to the digital interfaces of the subsystem. Transaction-level interfaces are defined for access to control elements of the SoC, such as the processor. The definition of these interfaces allows for interface coverage and transaction debug analysis to be used in developing the final implementation-level representation of the design.

To obtain optimal performance from the implementation, engineers develop the subsystem in an RTL format similar to the control digital interface. In these cases, structural assertions and coverage analysis are performed in a similar manner for a control digital subsystem.

Tests developed at the algorithmic and fixed-point level run significantly slower as the level of abstraction is moved down to the implementation level. Acceleration speeds the execution of these tests and provides for integration testing with larger subsystems. As part of the development process, blocks are put through a hardening process, where they are run through the mapping and synthesis stages of the hardware accelerator or the FPGA that will be used for system verification. Preparing the design for acceleration early in the development process eases the use of acceleration at the integration and system verification stages.

STEP 4: INTEGRATION AND DESIGN HARDENING

Once verification of the subsystem is complete, it is ready to be integrated into the final system. This integration is facilitated in the UVM through the FVP. Integration of the subsystem into the FVP is performed in a similar manner as the control digital subsystem discussed in Chapter 8. The main difference occurs when the algorithmic-digital subsystem needs to interface to an analog subsystem. Interfacing the algorithmic subsystem to transaction-level models is done using transaction interfaces similar to the control digital subsystem. Interfacing to analog or RF subsystems is done through converters. These converters play a similar role as a digital-to-analog (DAC) converter or analog-to-digital (ADC) converter plays in the final implementation. The converters convert the digital signal-level interface of the algorithmic subsystem to the continuous-time domain of the analog or RF models, as shown in Figure 35.

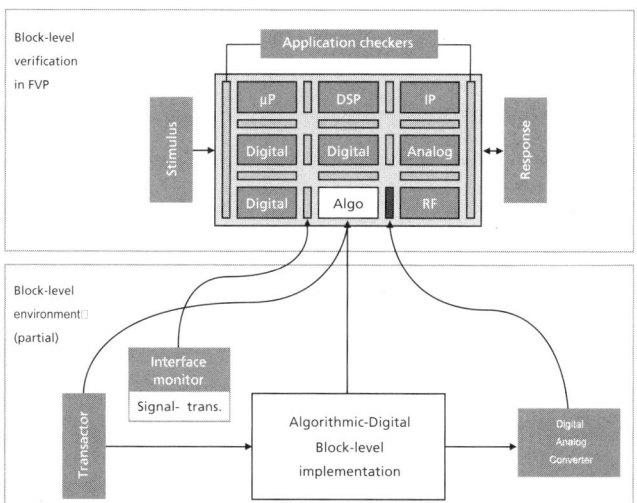

Figure 35. Integration of an Algorithmic Subsystem into an FVP

Large design blocks need the performance of a hardware accelerator to run long test sequences or to test multiple blocks integrated together. Once the block is stable, it is hardened by running it in isolation through the mapping process on the accelerator. This allows for the early detection of mapping or library issues that could stall the later use of acceleration. In a similar manner, the design is run through any synthesis or mapping processes needed for FPGAs or hardware prototypes used in system verification.

Chapter 10

Analog/RF Subsystems
Verifying analog subsystems

Analog subsystems include the classic analog designs as well as RF designs and high-speed digital designs developed in a full custom manner. The common thread in each design type is the need for functional verification at the logical level as well as the transient or AC level. Standard digital design verification separates the functional verification of the design down to the Boolean or gate level from the implementation verification of the gates and transistors. Verifying analog subsystems is more complex because functionality is equally impacted by the logical and physical design. Small changes in placement, component sizing, or silicon process can dramatically impact functionality.

The close relationship of physical design and verification with the functional verification of analog subsystems has led to an integrated approach to developing these subsystems. Standard digital devices might be split into separate design and verification tasks, with specialists in each area. The design and verification of analog subsystems are considered integrated tasks often performed by the same individual. Many analog developers consider functional verification as part of the larger design task. The UVM focuses on functional verification, but we cannot simply ignore analog subsystems as a design-only concentration. Most SoC designs being developed today contain both analog and digital subsystems. Verifying the individual subsystems as well as the integration of the subsystems is part of an overall UVM. So instead, we will focus on how the UVM connects to an advanced custom design process to provide successful SoC verification. This chapter introduces the Cadence Advanced Custom Design methodology and describes how it integrates with the UVM.[1]

CADENCE ACD METHODOLOGY

Cadence developed the Advanced Custom Design (ACD) methodology in 2003 to address the challenges of creating advanced custom/mixed-signal

---

[1] This chapter is taken from the Cadence Design Systems' white paper "The Advanced Custom Design Methodology," written by Kurt Thompson in September 2003.

designs. The methodology addresses the primary challenge of predictability by maximizing speed and silicon accuracy throughout the design process.

The ACD methodology is targeted at designers of full-custom designs, including those integrating digital standard cells with full-custom designs. The design scope focuses on key design domains of analog, custom digital, and RF, and supports their integration with digital standard cell blocks where integration is performed with a full-custom focus. The methodology is represented in Figure 36.

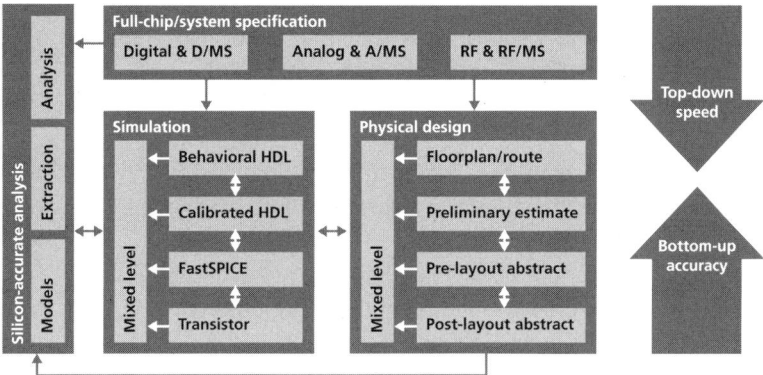

Figure 36.   Meet-in-the-Middle Approach

Predictability is predicated on two primary concerns: meeting the schedule from the beginning of the design process and meeting performance requirements.

Meeting the schedule requires a fast design process that supports thorough and complete simulation and physical design. The design process consists of numerous tasks. Many of today's chips contain multiple blocks from multiple design domains. Thus, it is imperative to design in as many blocks and perform as many tasks as possible in parallel, leveraging as much of the top-level IP throughout the process as possible. This leads to the concept of design evolution, where all the design IP is leveraged as it matures through the design process. The top-down design process when applied to both simulation and physical design facilitates a fast design process.

Multiple abstraction levels, from high-level design through the detailed transistor level, are combined to support a mixed-level approach that targets detailed design to only the points needed for a given test. This also allows for leveraging the top level and using that information for block design and subsequently reverifying the blocks in the top-level context.

At the other end of the spectrum is the need for silicon accuracy to achieve the required design performance. Silicon accuracy relies on base design data, such as device models supporting accurate simulation and technology files supporting interconnect and physical verification and analysis. Test chips, which often comprise critical structures known in the past to be highly sensitive, are also used to verify the feasibility of a process and the accuracy of its corresponding process design kit (PDK). Often, a design group needs to add components to the PDK to support a particular design style. Device models might need to be expanded to combine or add corners, statistical modeling, or other approaches the design team needs.

The silicon accuracy data is driven through the design process by detailed transistor-level analysis, including layout extraction. These make up the lower level of the abstraction chain, which then supports the calibration of these results to higher levels of abstraction. This is the bottom-up design portion of the ACD methodology.

The top-down and bottom-up processes work in parallel, producing a meet-in-the-middle approach that balances the need for speed through the design process and silicon accuracy, ultimately producing a predictable schedule and first-pass success.

THE MEET-IN-THE-MIDDLE APPROACH

The ACD methodology relies on a meet-in-the-middle approach as the most pragmatic method to achieving predictability on complex designs. This is accomplished by leveraging the fast capabilities of top-down design together with the silicon accuracy capabilities of bottom-up design. These two primary vectors combine and essentially merge where the majority of the design activity cannot be described as either top-down or bottom-up, but as a combination of the two.

Multiple abstraction levels are used to represent the evolution of each piece of the design. In simulation, behavioral models, which grow more detailed as the design process moves forward, are used initially to bring in measurements and data from post-layout analysis. In physical design, initial size estimates and initial block abstracts are updated as more design information becomes available, to where the actual layout is used for the top level. The designer is actually working in the middle most of the time, with some blocks at the fast, top-down stage, and some annotated with additional design data and silicon accuracy using the bottom-up process.

Dealing effectively with legacy IP is often the factor that forces a meet-in-the-middle approach. Rarely does a design team start from a clean slate (in these rare cases, a pure top-down methodology can be employed from the beginning). Legacy IP blocks must be upgraded to support the ACD methodology, which is done using a bottom-up approach. In most cases, the block only has the transistor level and layout abstraction levels supported. As a result, the abstraction levels are derived bottom-up and then fed to the top-down process.

Abstraction levels serve as the foundation of the meet-in-the-middle approach. Both simulation and physical design have predefined abstraction levels, which are updated through the design process and support the mixed-level capability. The abstraction levels are:

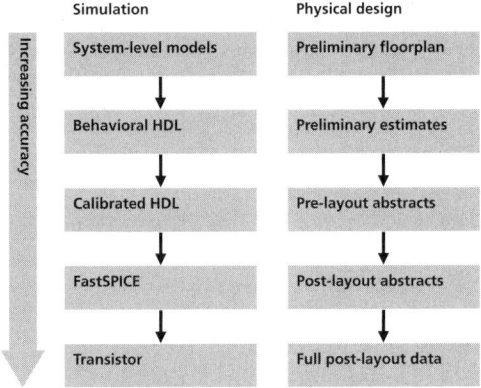

Figure 37. ACD Abstraction

- System models (simulation)—Generated from the FVP. They are the highest level of abstraction represented. This also includes test-benches for system simulations.

- Behavioral HDL (simulation)—Most often refers to Verilog, Verilog-AMS, Verilog-A, VHDL, VHDL/AMS, or VHDL-A descriptions. At this level, only the circuit functionality is described, and the models are targeted for fast run time.

- Calibrated HDL (simulation)—HDL models are calibrated off of transistor-level simulations, making the initial behavioral HDL models more accurate representations of the circuit behavior.

- FastSPICE (simulation)—Uses the same transistor-level descriptions as SPICE. Running a FastSPICE simulator gives the designer silicon accuracy versus fast run-time options. Therefore, the FastSPICE option can be considered a separate abstraction level.

- Transistor (simulation)—SPICE-level simulation at the most accurate level.

- Preliminary floorplan (physical design)—Highest level abstraction for the physical design process. At this stage, relative placement, initial pin optimization, and other floorplanning investigations are supported.

- Preliminary size estimates (physical design)—These are based on previous experience, information from derivatives, initial process feasibility studies, or any information on which the design can base a block size.

- Pre-layout abstracts (physical design)—When a transistor-level description is ready, the preliminary size estimate can be updated to more accurately reflect the layout, prior to the layout being completed.

- Post-layout abstracts (physical design)—With layout complete, a final abstract matching the physical representation can be provided to the routing process.

- Full post-layout data (physical design)—Supports final physical verification as well as chip finishing tasks and final tapeout.

The ACD methodology then uses these abstraction levels, which serve as its components, across the entire design process. How these abstractions are put together and used determines the level of predictability for the design. The design team must manage a plan up front to determine where to bias the process for silicon accuracy or speed, define a mixed-level definition to accommodate it, and execute the meet-in-the-middle process.

THE ACD FLOW

The ACD methodology can be described through a task-based flow as shown in Figure 38.

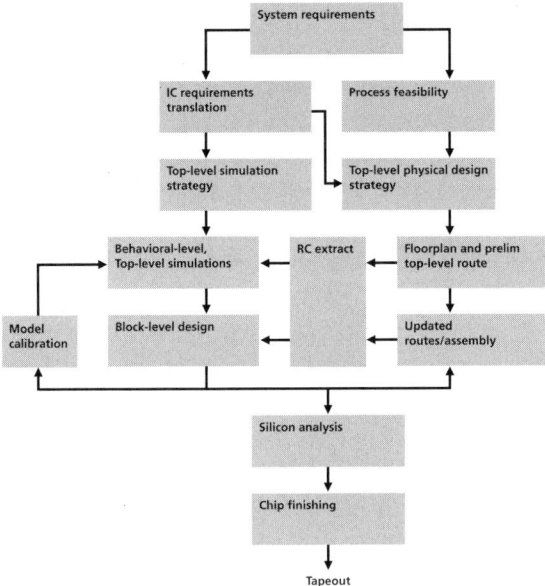

Figure 38.   ACD Flow

A fast, silicon-accurate design process is achieved by working on the final design up front and early in the design process. Since the top-level tasks pose the most risk, and saving them until the end of the design process invariably produces delays and iterations, it is imperative to move these tasks up front in the design process. As the process moves along, blocks get further defined and fed into a top-level evolution where top-level tasks, simulation, and physical design are continuously verified with updated design collateral as these pieces mature. Supporting and maintaining this evolution is what ensures predictability.

The primary advantage of enforcing a methodology that lets the top-level continuously evolve is that difficult tasks, such as silicon analysis, RC extraction, routing, and physical verification, can be performed early on. While this is not the final version of these tasks, interim data is used from these tasks early on to drive design tasks through the hierarchy and support a fast design process. Also, because these tasks are performed early on ensures that they can be repeated downstream as the design matures. Knowing how long top-level route through verification takes per design iteration helps the design team predict the time needed for each stage in the design process more confidently.

System Requirements

System requirements is the first task in the flow. This is where the UVM and ACD methodologies come together. The development of an FVP at the architectural development stage of the UVM provides the system requirements that are fed into the analog subsystem stages. The FVP can model analog, RF, or custom digital blocks at a high system level using C-like algorithms wrapped in a transaction-level wrapper. The FVP is given to the analog subsystem teams. The team provides models for the individual subsystem functionality and a system-level test environment. They can begin the fast top-down simulation process using these models and the FVP environment.

Through the ACD flow, the functionality can be modified or refined. These changes are updated in the FVP model and provided to the system-level verification team for analysis and distribution to other subsystem teams. The meet-in-the-middle process develops more accurate models. Where it makes sense, these models can be integrated into the FVP models and tested within the system-level test suite. Each of these steps ensures that the verification and integration of digital and analog subsystems are performed in a unified manner.

Process Feasibility

With IC requirements generated from system specifications and the FVP, process technology selection must occur. Evaluations of silicon accuracy capabilities and various integration strategies must be performed to verify the feasibility of the proposed integration approach. Issues such as performance, noise characteristics, cost, circuit type, and risk are all considered.

IC Requirements Translation

The system design process produces specifications that the IC must meet. The system design process leverages the UVM in using these requirements through system-level models, testbenches, and measurements. The testbenches may be further enhanced to match specific IC specifications where the specification-driven environment can be set up. The specification-driven environment then drives the chip level. Subsequently, the block level tests in a manner consistent with the original requirements given to the design team.

Simulation Strategy

With a process selected and its feasibility and silicon accuracy ensured, the strategy by which the design will be built can be defined. At this point, the design team has made primary decisions as to the integration strategy of the design and identified the constraints to insert through the design process based on silicon accuracy data.

Successfully executing a complex design is contingent on the thoroughness of the planning up front. No design can come together smoothly by accident. With a strong plan in the beginning that specifies the top-level and block-level requirements and the mixed-level strategies to use, a meet-in-the-middle approach can drive each block design to ensure full coverage of important design specifications and smoothly allow for blocks to have different schedule constraints. By using the most up-to-date information available at any given time, blocks that are done earlier can be verified in the top-level context and be ready to go. This enables time and resources to focus on the more complex blocks, which can also be using the most up-to-date information.

At this point, the high risk points flagged for targeted verification are examined. These could be areas such as analog/digital interfaces, timing constraints, or signal/data paths. What is extremely important at this stage is to look at a simulation and physical design approach that can support verifying these risk points. The mixed-level approach needs to be examined to determine the abstraction level these points are described at. For example, a key analog/digital interface might need both the digital interface and analog interface sections described at the transistor level, with detailed parasitic information in between to ensure bit errors do not occur. If this is the case, it should be determined how the design will be partitioned to allow this simulation to occur in an efficient and repeatable manner. Often, this interface can only be meaningfully tested at chip level over a variety of simulation setups. Predictability is predicated on the assumption that all critical items are part of a simulation and verification strategy and are repeatable and reliably execute throughout the design process.

With critical circuit issues identified, the next step is to tackle design partitioning as part of the simulation and physical design plans. It is important to consider design partitioning from a functional perspective as well as an enabler to use the design tools effectively to verify the identified critical circuit issues. The designer must consider the ability of the tools to handle certain types of analysis, and design the circuit hierarchy to isolate each issue and efficiently tackle the problems associated with it.

Design partitioning is nearly always looked at from a functional perspective. It is natural to partition in this way because it leads to block specifications and layout partitions, which in turn lead to a natural top-level simulation strategy. It is important to keep this functional partition intact. However, as in the case of an analog/digital interface, you must also consider how the mixed-level capability can be employed to verify this interface at the top level. One approach is to ensure that the block partitioning on the analog side has an interface piece that can be swapped at transistor, and that the digital section also has its interface piece that can be swapped at transistor. Parasitics can be added inside these transistor sections, and interconnect parasitics can be back-annotated in between these blocks. The rest of the chip level can then be described at the HDL level of choice for increased simulation speed. This is represented in Figure 39.

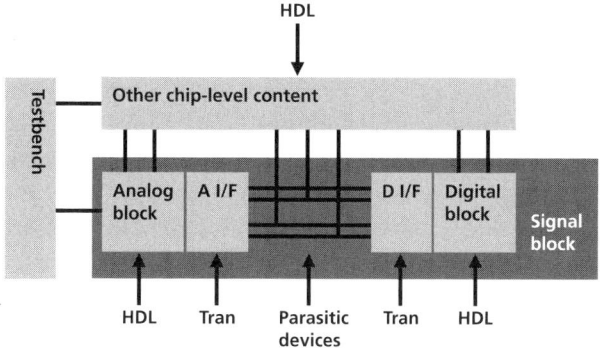

Figure 39. ACD Partitioning

If the design partitioning does not take this situation into account, the next option is to bring the analog and digital blocks into transistor level (assuming this interface is critical and needs to be simulated at the transistor level). While this achieves the objectives, it is quite possible this simulation would be quite slow regardless of which simulator was used. Waiting for transistor level also requires that the transistor level is complete, while the partitioning approach allows for the analog and digital sections to be completed at their own pace. If the interface sections get done first, the interface itself can be tested before the analog and digital core pieces are complete, aiding a fast design process. The ability to simulate the interface of concern at transistor level satisfies the silicon accuracy requirement. As the design evolution occurs, it might be desirable to bring more pieces into transistor level or to simulate the analog blocks with the analog interface in transistor. This adds to the predictability of the design process by enabling evolution and resolving critical design issues early on.

Thus, the simulation strategy must be comprehensive to account for all tests that must be performed and ensure that the design database is partitioned conducive to that strategy. The simulation strategy should also take into account the completion estimates of each individual block and specify the mixed level for that simulation. For example, the following table lists some example sections of a simulation strategy.

Table 3.   Example Portion of Simulation Strategy

Top-Level Test	Testbench	ADC	DAC	DSP	CODEC
Codec Verification	Functional A	Verilog-AMS	Verilog-AMS	Verilog	SPICE
BER Function	System BER	Verilog-AMS	Verilog-AMS	Verilog	Verilog-AMS
DSP Verification	Functional A	Verilog-AMS	Verilog-AMS	FastSPICE	Verilog-AMS

For large SoCs, separate tables may be necessary for the major blocks. Often, the first level of hierarchy for each block is much like a large chip and can have all the issues associated with a chip. In these cases, separate add-on tables, such as the one above, might exist for each block at the top level and subsequently through the hierarchy, where applicable.

As the design evolves, analog HDL descriptions can get more accurate as transistor-level simulation results are back-annotated into the models. There is some simulation speed price for this. The simulation strategy is amended where accurate models are needed. In block cases, it is likely that accurate HDL is used across the board in conjunction with FastSPICE capability, and SPICE-level capability for the most sensitive circuits.

For complex blocks that require some silicon accuracy at the top level, the block designer might specify a particular mixed-level configuration when simulating at the top level. At the top level, this block-specific configuration exercises the simulation strategy. One view might be a non-hierarchical behavioral view for the block, another might contain the internal sub-blocks at accurate-HDL or transistor levels. This hierarchy and configuration must be managed to match the simulation strategy.

Behavioral-Level Top-Level Simulations

The fast top-down design process necessitates a top-level HDL description of the design. This description is consistent with the partitioning specified through the simulation strategy and follows the declared hierarchy. The simulations performed are consistent with the specification-driven environment specified above, where individual tests are documents in the simulation strategy. These simulations are then used as test beds for blocks under test. Block-level testbenches are derived from the chip-level simulations capturing block-level stimulus.

Block-Level Design

Block-level design is based initially on the top-level simulations that verify the block specifications. Block-level design then encompasses the detailed, silicon-accurate, transistor-level design. This also includes incorporating parasitic data and performing silicon analysis.

Model Calibration

The silicon accuracy process, enabled through the bottom-up flow, requires higher level abstractions to maintain as much of the silicon-accurate information of the individual blocks as possible. This requirement is met through calibrating functionally correct behavioral models with silicon-accurate design data derived through the post-layout transistor-level simulations.

Physical Design Strategy

The top-level physical design strategy is specified in parallel with the simulation strategy, although there is some dependence on the specification of the hierarchy and partitioning from the simulation strategy. The purpose of the physical design strategy is to look at routing constraints, floorplanning constraints, and initial placement based on block characteristics. Based on these constraints, decisions, such as at which level top-level routing is performed, are flagged.

Floorplan and Preliminary Top-Level Route

The floorplan and preliminary top-level route are critical in supporting a fast design process. When an initial route has been completed, a repeatable process exists to support continuous design evolution. The setups and constraints are reused and modified as the design evolves. These setups identify issues at the top level early, where both design and tool issues can be fixed before tapeout. Predictability can only be achieved if these steps are done early and repeated throughout the design process.

Updated Routes

As the design evolves, the initial setups are used to route updated physical abstracts that represent more accurate size estimations, ultimately through accurate block abstracts generated from the completed block layout process.

RC Extraction

Whenever possible, post-layout analysis on the first cut database should be set up, even if the results are not totally meaningful at this point.

Silicon Analysis

At the top level, silicon-accurate analysis that functional-based simulation does not catch is performed. This includes IR drop, electromigration (EM), and substrate noise analysis. Some silicon analysis can be performed at the block level, and some can be performed during the updated routing tasks.

Chip Finishing

Chip finishing includes tapeout preparation tasks, such as adding a PG test, layer editing, adding copyright and logos, and metal fill. At this point, it might also be necessary to make final edits based on last minute design needs.

Chapter 11

Integration and System Verification
Verifying system operation

The final stages of the UVM are system integration and verification. Once each of the individual subsystems has been verified using the UVM, it is time to bring them together and verify the operation of the system as a whole. System integration and verification is where a verification methodology is put to the test. Fragmented verification approaches fall apart when you try to integrate incompatible test environments that have been developed in complete isolation. Using a unified methodology facilitates efficient integration utilizing testbench reuse and common models. After integration, final system verification is performed to ensure correct operation under real-world environments. System verification techniques can vary depending on the specific application, but the goals should remain the same. This chapter focuses on best practices and techniques used in system integration and verification.[1]

SYSTEM INTEGRATION

In the UVM, each subsystem is continuously verified using the FVP as a common reference, so the integration and test of the system should be straightforward. The SoC team integrates each implementation block into the FVP one at a time and runs the system test suite to verify the integration. The lower level assertions and monitors should also be included in the integration testing to aid debugging. The test plan is run with the FVP for comparison checking. Once the system has been verified as equivalent to the FVP, the implementation is considered the implementation-level FVP, and the original FVP is the transaction-level FVP. The design is then ready for system verification.

[1] Parts of this chapter are taken from "Hardware-Based Verification Is Necessary for Today's Million Gate+ Designs," by Ray Turner, Jr., published in Cadence Design Systems' *Verification Talk* newsletter, March 2002.

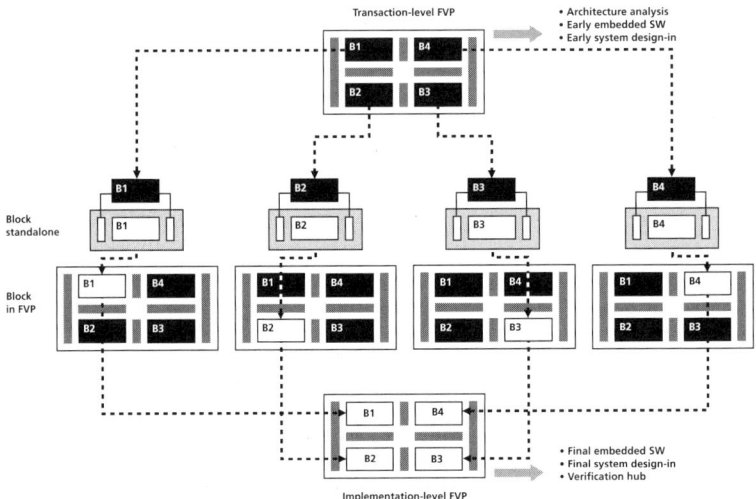

Figure 40.   System Integration

Integrating a Subsystem into the FVP

Verifying the individual subsystem implementations in isolation in the FVP does not verify that two subsystem implementations will work together correctly. The SoC team must integrate each subsystem implementation together in an organized fashion to verify that these implementations function together correctly. The SoC team does this testing by adding one implementation block at a time into the FVP, separating implementation-level blocks from transaction-level blocks with transactors. Subsystems are added one at a time, if possible, to monitor the possible causes of integration errors. Once a subsystem is verified to be operating correctly with its surrounding implementations, performance is improved by putting that subsystem into hardware acceleration for subsequent integration of other blocks.

An important consideration for the SoC integration team is when to move to a pure signal-level, top-level interconnect. The final SoC implementation will have a pure signal-level top level, which is usually provided by either the physical design team or the system design-in team. These top levels can consist of tens of thousands of individual signals for large SoCs. The only time these top levels are verified is with the final integration, so it is imperative that the SoC team provides an efficient method for this verification. The SoC team should move to the physical signal-level top-level interconnect as soon as possible after the transaction-level FVP has been verified. This might mean keeping two top-level interconnect models in sync for a period of time during the transfer. The signal-level top level is also necessary, since it is common for many side-band signals to appear between implementation blocks that are not necessary in the transaction level.

The SoC team makes the integration process smoother by verifying the physical signal-level top level with the transaction models as shown in Figure 41. In this example, transactors are placed at the interfaces of each model, and interface monitors are converted to signal level and placed between pairs of transactors. The test suite is then rerun to verify the signal-level top level. The transactors and interface monitors, which will be reused by the subsystem teams, are also verified in this configuration.

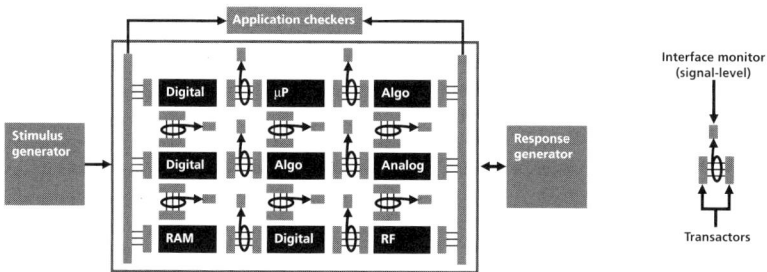

Figure 41.   FVP Signal-Level Interface

Simulation Acceleration

Hardware-based verification provides several modes of simulation acceleration, with varying levels of performance improvement over simulation alone. The first mode is accelerated co-simulation, which is the easiest mode to implement. With accelerated co-simulation, also known as lock-step co-simulation, engineers compile and download their design on a dedicated hardware verification engine while leaving their behavioral or C++ testbench in the simulation environment. With the simulator or C-testbench within the workstation communicating in lock-step fashion with the design in the accelerator, simulation performance is increased from two to ten times faster than traditional software simulation. This is primarily because the bottleneck—the workstation—is now only responsible for a fraction of the total simulation load—the testbench.

Lock-step co-simulation is easiest to implement because it involves little change to the simulation environment. In most cases, the design is simply split into two pieces: the behavioral portion, which is generally the testbench, remains on the workstation, while the synthesizable portion, along with memory constructs, are loaded onto the accelerator. Beyond this, little or no change is made to the environment, which allows for fast implementation. This approach has some drawbacks, which are primarily caused by the fact that most testbenches include hundreds of signals that must communicate with the design on a clock-by-clock basis and, in many instances, several times per clock. This high level of communication limits the overall performance of the co-simulation environment, since the hardware system, which is able to execute the design many thousand times faster than the simulator can execute the testbench, must wait after each and every clock tick for the testbench to complete its execution.

The next level in performance is accelerated, transaction-based co-simulation mode, which can be one hundred to one thousand times faster than traditional simulation. This is a new mode that overcomes many of the limitations introduced by the relatively slow workstation. Rather than communicating on a clock-by-clock basis, a transactional testbench introduces a high-level protocol for communication between the workstation and the accelerator. This protocol, in combination with a specialized, low-latency I/O channel, allows data and commands to be sent between the simulator and accelerator at high speed and in parallel. Once the data and high-level command are received by the accelerator, it is free to stimulate the design at full speed, and only needs to communicate with the workstation once the given task is complete. The implementation of transaction-based co-simulation requires that a small portion of the testbench, known as a transactor, be synthesizable. For many design environments, the small amount of additional work to create the synthesizable transactor is well worth the substantial improvement in overall co-simulation performance. Even with transaction-based co-simulation, the testbench remains a bottleneck to overall performance.

Hardware's fastest mode—synthesizable testbench mode—is a hundred times faster than accelerated co-simulation, boosting simulation run-time performance by up to one hundred thousand times. This high level of performance is achieved by eliminating the workstation all together by loading the entire testbench onto the accelerator along with the design. Without the workstation as a bottleneck, the overall verification performance is maximized and runs at the same speed as in-circuit emulation—typically 300k to 750kHz. Though this mode delivers the fastest performance, it does require that users supply a synthesizable testbench, which takes additional effort. Some users, like those designing CPUs and networking chips, find this worthwhile, since it delivers the ultimate level of performance. Many customers initially implement lock-step co-simulation for their first accelerated project and later move to transaction-based co-simulation, synthesizable testbench acceleration, or both.

Order of Integration

The order of integration depends on the design and the delivery time of each piece. If the design contains analog connected to a digital signal processor (DSP) block, with the DSP block connected to a control-based digital block, the analog and DSP blocks are verified first, and then verified with the control-based digital. If the design contains analog blocks directly connected to control-based digital along with algorithmic-based digital, the control-based digital is integrated first with the analog independently, and then with the algorithmic-based digital independently. When these two integrations have been verified, the system as a whole is verified.

SYSTEM VERIFICATION

The goal of the system verification phase is to verify the system under real-world operating conditions. The UVM utilizes system verification for several roles. First, to verify that the testbench environment used to stimulate and check the implementation has accurately reflected the system. It also provides a mechanism for hardware-software co-verification in a realistic environment. Up this point, software development has been done on a model or Instruction Set Simulator (ISS) attached to the FVP or an implementation model using only basic software debug tools. In system verification, software is run either on the actual CPU or a mapped version of the CPU utilizing all the software debug tools available in a real-world environment. System verification is also used as a design chain handoff mechanism, allowing early access of the implementation to design chain partners.

Three basic types of system verification methods are used in the UVM: software-based simulation, hardware prototype, and emulation. Which method to use depends on the application and the skill set of the team. Software simulation works well for smaller designs that do not need to run at fast speed for long periods of time. Setup and conversion to software simulation methods is straightforward. FPGA development platforms work well for modular designs such as SoCs. However, setup and conversion can be cumbersome for someone not experienced with FPGA development and partitioning. The FPGA solution can run much closer to the speeds of the system, but it provides only limited visibility to help debug any problems encountered. Emulation systems work well for large designs that do not need to run at full speeds, but do need to be accelerated much faster than simulation. Emulation systems interface well to external devices and provide excellent visibility and support for debugging design issues.

The UVM speeds up system verification in several ways. Reusing models and testbench elements shortens development time. Reusing transaction-based models and interfaces improves the speed of the system, thereby decreasing test time. Reusing assertions and using a common user interface from the implementation stages speeds debugging. Finally, the hardening process of digital subsystems decreases the time to working emulation or prototype.

Software-based Simulation

Software-based system verification provides the greatest amount of observability and the smallest modification time of the three methods. However, performance can limit the amount of verification performed. The first step is modifying the testbench. The implementation is now controlled and monitored by the actual external environment. The testbench is modified to remove stimulus generators. Response checkers are modified to become passive monitors for debugging. Hardware-dependent software must be loaded into the processor model in the simulation or controlled through an instruction set simulator. If the system contains mixed-signal subsystems, such as algorithmic digital or analog subsystems, they are either modeled in a higher level of abstraction for simulation, such as a TLM, or black-boxed and ignored.

Stimulus can be provided in several different ways:

- Interfacing to test equipment and capturing the stimulus for playback on the implementation
- Using API interfaces to the simulator to receive and drive data to and from a workstation or network
- A model or TLM that mimics the real-world stimulus

The output of the system is verified through comparison and analysis. It can be compared for accuracy to an existing system or the FVP. Analysis can verify user interfaces and performance requirements.

Advanced verification techniques continue to be used throughout system verification. All the assertions defined at the architectural, interface, and structural levels are reused to aid debugging and detecting errors found with the new stimulus. Architectural-level coverage goals can be reverified to ensure that stimulus is working correctly, but lower level coverage monitoring is not used. Transaction-level interfaces provide the same debugging environments used in subsystem development and integration. Acceleration increases the simulation speed of the design in a similar manner as with system integration.

Hardware Prototypes

Hardware prototypes are hardware systems built to replicate the real system environment using programmable hardware, such as FPGAs, to represent the implementation. Hardware prototypes might provide the most high-performance solution for system verification, but also require the most work. The process begins with developing the prototype system. In most cases, the prototype system must be built or modified for system verification use. The prototype system requires a tested board with standard interface components, along with observation interfaces. The digital implementation is placed in programmable hardware. The analog and mixed-signal subsystems are implemented on the board.

Once the system is built and debugged, the design is compiled into programmable hardware devices, and system clock speeds are chosen to meet the timing of the compiled implementation. The subsystem teams might have already verified that the blocks map correctly into programmable hardware as part of the block-hardening process. Stimulus is provided through the prototype system board and can be driven through test equipment, external workstations, or networks. Software loading and debugging are done in the same manner as the real system. Service processors can be used to boot the system, and software debuggers connected through JTAG interfaces can be used to debug the software. The response checking is done in a manner similar to the real system. Response to the stimulus can be captured for analysis by test equipment, or the application can be tested to the user requirements.

Advanced verification techniques are not commonly used with hardware prototypes. Assertions and coverage monitors do not map easily or have a common interface to standard programmable hardware devices. Hardware prototypes can be replicated for design chain partners, providing early access to the implementation-level FVP. This requires a great deal of up-front planning and back-end support.

Emulation

In-circuit emulation provides the highest run-time performance for regression testing, hardware-software co-verification, and system-level verification. In-circuit emulation replaces the testbench with physical hardware, which is typically the system for which the IC is being designed. Working at the system level, in-circuit emulation verifies the IC as it interacts with the system, which includes system firmware and software. Rather than using testbench-generated stimulus, which is often limited in scope, in-circuit emulation is able to use live data generated in a real-world environment. Data generated at high speed, using industry-standard test equipment, is also available with in-circuit emulation. In many cases, the last handful of corner-case bugs, which if undetected would result in costly chip respins, can only be discovered through the interaction of the IC in the context of the system, with software, firmware, and live data.

In-circuit emulation also bridges the debug environment between simulation and physical hardware. Even when running in-circuit with a live target system, emulation provides a comprehensive debug environment with full visibility into the design being emulated. Combined with very fast compile times (typically 4 million IC gates per hour on a single workstation), in-circuit emulation becomes similar to simulation, where bugs can be found quickly, fixed, and recompiled, often in less than an hour.

Several different applications of in-circuit emulation are available. Two of the most commonly used include vector regression and hardware/software co-verification. Using the emulation system in vector regression mode enables users to run their sign-off suite of vectors at high speeds, which is valuable for final certification of any design before tapeout. With vector regression mode, test vectors are loaded onto the emulator along with the design. These vectors are then used to stimulate the design, with the output vectors being captured. For long regression tests or suites of tests, additional vectors can be loaded onto the emulator as the previously loaded set of vectors is executed. Likewise, results from the previous set of test vectors are off-loaded and stored on disk as the current set of results is captured. This ability to load and unload one test while another test is executing maximizes throughput of the emulator.

To create a chip tester-like environment, an emulator can optionally compare the vector test results with "golden result vectors" on-the-fly and report pass/fail results. For failed tests, which can be debugged later, vector mismatches are highlighted in the waveform display browser. With vector regression mode, the emulator can be kept constantly testing designs, which dramatically reduces the time required to complete a large test or entire regression suite, often from weeks to hours.

Hardware-software co-verification is a powerful option to in-circuit emulation that can dramatically reduce the verification time and development time of today's designs. By providing a functional system environment, an in-circuit emulation system can be used to develop system-level software, even as the IC design is being verified. By developing software in parallel with hardware, not only is the development schedule effectively compressed, but the system-level software becomes available as an additional verification tool for the hardware, which provides another means to uncover deeply hidden bugs. By testing software while the hardware is still being developed, changes can be made in the hardware design before tapeout to yield optimal solutions.

To provide a familiar software development environment, both hard- and soft-IP can be interfaced to any microprocessor ICE or RTOS debugger for early software validation and debug of an emulated design. Soft-IP can be modeled within the verification system along with the IC, while hard IP can be interfaced to the system using the in-circuit interfacing logic.

For interfacing an emulated design to devices that are speed sensitive, there are a variety of specialized verification environments. These environments can be used to interface an emulated design with full-speed interfaces, such as Ethernet, audio, video, PCI/X, and 3G wireless. These sources of live data enable the verification of hardware/software interactions in real-world environments.

By moving from the context of the IC to the context of the system, emulation delivers the ability and performance required to boot operating systems, develop and test device firmware and drivers, and even interact with running applications, all while providing a fully featured debug environment.

SECTION 3
TOOLS OF THE TRADE

Chapter 12

System-Level Design
System modeling, software, and abstraction

Early system modeling and verification are the cornerstone of a modern unified verification methodology. Earlier we introduced the FVP and showed how it can unify a verification methodology to improve speed and efficiency. An FVP is one example of a system model that has been used for many years to assist in system design, system verification, and software development tasks. In this chapter, we will look at the issues that can be addressed with an FVP and how software development and functional verification can be unified in a methodology. We will also address the important topics of design and verification abstraction.

ISSUES ADDRESSED WITH AN FVP

The system model addresses many of the most difficult issues verification teams face today. One of the most important issues is incomplete or incorrect communication of architectural and design information. Architecture and design information passes among groups or engineers in many ways. The most commonly used format is a written specification but, unfortunately, it often lacks all the necessary information, has ambiguous information, or might not be kept up as changes occur. These deficiencies often lead to bugs being introduced into the design and slipping through the verification process.

A system model can address these communication problems because when it returns a non-ambiguous result to specific stimulus, it creates an executable specification. The architects or system designers can specify and verify the exact operation of the system in a system model. This system model can then be provided to implementation, integration, and verification teams. Each of these teams designs or verifies their part of the system to meet the operation of the system model. Verification can compare the response of the implementation to the response of the system model. If the responses do not match, either the design is in error or the specification is incorrect. Either way, bugs are identified and resolved faster using the system model as an executable specification.

Some bugs found during the verification process are not due to the implementation of the design but are in the original architecture. These bugs can result from incorrect assumptions made by the architects or system designers and often require large portions of the design to be redesigned. Finding these architectural bugs as soon as possible limits the amount of rework by designers and verification. The system model can help find bugs that have slipped through the verification process, and it can help find bugs sooner. Developing a system model early in the development process allows the verification team to verify the architecture and system design before implementation begins. Verifying the system model early identifies bugs sooner than waiting for the implementation to be completed. Correcting these bugs early saves implementation and verification time and resources.

In addition to finding bugs, verification teams face efficiency and productivity issues. Many development teams view design and verification as serial processes. They believe that verification does not begin until the implementation has been completed, because you need to have something to test before you can begin testing. Many verification tasks, such as testbench development and test writing, can be done in parallel with the development of the implementation, but the verification engineers cannot test or debug their testbench or tests until the design is made available. When a system model is available, it can be used to test the verification environment and to develop and debug the verification tests and infrastructure before the implementation is completed. This means that once the design is made available time is not lost integrating, debugging, and bringing up the verification environment. This can dramatically improve the verification schedule and spread the number of resources more evenly.

The system model can also be reused as part of the verification environment, thereby saving testbench development time. A large portion of the development in most testbenches is checking the response of the design to stimulus and determining whether it is correct. If the system model is developed as an executable specification, the system model can be used to predict the correct response for the design. This saves time in developing testbench components. The system model can also be used as a verification environment. If the system model is partitioned correctly, it might be possible to replace parts of the model with the actual implementation and reverify the model. Verifying parts in this manner can facilitate reusing the system-level tests to verify the operation of the part within the context of the entire system. This reduces the amount of tests that need to be written.

If the design is part of a design chain, the design customer might need access to the design early in the development process. The verification team is often required to provide this prototype. Many teams deliver an implementation-level model or a prototype in the form of an FPGA. Instead, a system model of the implementation can be delivered earlier in the design process and in a more useable format for the customer. The customer might still require a more detailed prototype later in the process, but the system model can satisfy them until it is available.

VERIFICATION AND SOFTWARE DEVELOPMENT

Throughout this book we have spoken of the connection between functional verification and software development. Unifying these two domains could provide great strides in improving the speed and efficiency of the development process. While striving to unify system models and development environments is noble, it is important to realize the uniqueness of the individual tasks of functional verification and software development. These domains have different goals and utilize different processes. Attempting to develop one environment and set of processes for verification and software development would in many ways be counter to the culture of each group and result in an overall loss in the project's speed and efficiency.

Instead of focusing on unifying to a common environment, teams should focus on unifying the connection points between software development and functional verification. This allows each team to follow their own best practices while utilizing common deliverables. In this section, we will discuss the connection between functional verification and software development from the perspective of the functional verification engineer.

Advanced functional verification teams have two major concerns regarding software environments: how do you provide a common model that software can use, and how do you utilize a software environment in the verification process. In many cases, the software team is viewed as a customer for the functional verification team. The functional verification team must provide the software developers with an environment that allows them to develop and test their software before silicon is available.

The functional verification team should focus on delivering a fast, accurate, and flexible model of the design for the software team to use. The model must be fast enough for software developers to test code sequences that consume millions of cycles of simulation while providing response times suitable for interactive development and testing. The model must accurately reflect the implementation, including accurate memory maps and register definitions, it should provide realistic responses to software operations, and must be flexible enough to evolve as the implementation evolves. At first, a TLM may be suitable, then followed by various parts modeled in HDLs, and then possibly a hardware-accelerated model. The model and environment must meet the needs of the software team while providing enough visibility to allow for hardware debug should functional errors be found.

Using the System Software Environment for Verification

When using the system software environment to verify a design, the main goals should be to verify the intended implementation, verify the integration of software and hardware, and provide a realistic software environment. The environment should be able to stimulate and verify the implementation in much the same way as a traditional testbench, without developing the testbench infrastructure and tests. The environment should verify that the implementation in coordination with the intended software is operating correct. Finally, the environment should provide a real-world environment that verifies the design under similar conditions that it will face in normal operation.

Utilizing software environments for functionally verifying the design can be a fast and efficient solution, but it also has its drawbacks. A software-based environment provides fast and easy stimulus generation, but the stimulus can be difficult to control. Software usually uses a small subset of the functionality and does not stress important corner cases where bugs are often found. Even when stimulus can be precisely controlled, bugs can be missed due to a lack of adequate checking. Software environments lack visibility and often rely on register read-backs or memory compares to verify correct operation. Bugs that do not easily propagate to these checks can easily be missed. When detected, bugs can be difficult to find in software environments due to the lack of visibility.

Given the number of goals and caveats for using a system software environment for functional verification, it is not surprising that the teams doing this and the degree to which they do it is highly dependent on the application. Basic control systems that operate over a small range of variability can most fully utilize a software environment for functional verification. Complex custom hardware applications that rely on software for initialization, error handling, and background services tend to rely more on traditional testbenches and use software environments only at integration and final test. There are three basic types of applications that utilize software in the functional verification process.

Microcode Engines

Many systems utilize a microcode engine to provide some flexibility while maintaining custom hardware speeds. Graphics engines and network applications may implement protocols or algorithms in custom microcode and develop custom engines to process the limited instruction set. Verification of these engines must focus on verifying both the engine and the software. Verification is made easier because of the limited instruction set and limited number of applications. Unfortunately, the software is often not available when the system is being verified. In many cases, the engine must be verified to correctly run microcode that will not be written until after the silicon has shipped to a customer.

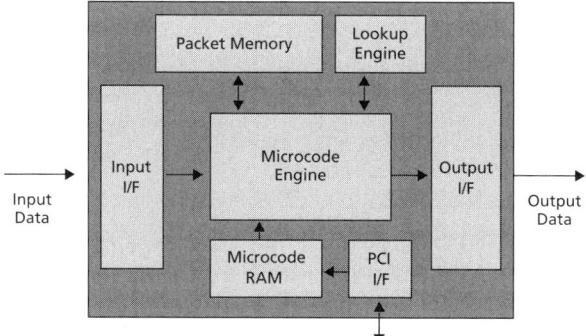

Figure 42.   Microcode Engine

Verification teams can use random code generators to verify these types of systems. These generators create code snippets that can be loaded into the engine and run. Usually, the generation must be tightly constrained to allow only valid code snippets. Software environments allow microcode engines to verify both the hardware and the integration of software.

Hardware Platforms

Hardware platforms provide a configurable environment for a wide range of possible system applications. These platforms are implemented as SoCs with one or more standard processors, memory, and basic application interface units, all connected to a standard common bus structure. These systems often use a standard preverified processor, so verification is focused on system integration and any new hardware blocks that may interface to the processor. The software that runs on hardware platforms has multiple layers, including low-level drivers, operating systems, and applications. The software is usually readily available, and verification only needs to focus at the device driver layer.

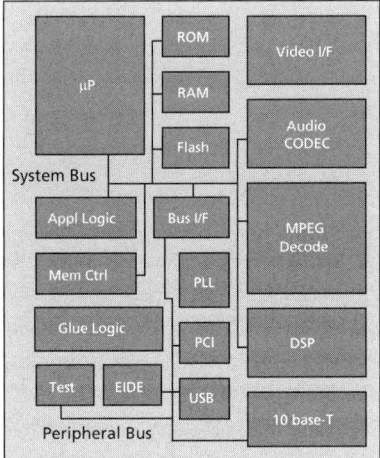

Figure 43. Hardware Platform

To verify these platforms, the processor is replaced with a bus functional model (BFM) or an ISS to improve simulation speed and provide the necessary visibility and control. Final verification is usually done with an implementation model of the processor to verify interconnect and integration. Advance verification teams utilize software environments for the verification of hardware platform SoCs to verify standard interfaces and configurability.

Software Algorithms

An algorithm can be implemented in hardware by converting it to a custom implementation or by running it on a standard processor. The algorithm is first developed and verified in its software form in an environment such as an FVP. Functional verification is left to verify the final implementation of the algorithm and the interface to the system. If the algorithm is implemented with custom hardware for speed or size reasons, verification focuses on verifying the conversion from algorithm to RTL or gates and the correct interface to the rest of the system. The verification team can utilize the original software algorithm as a reference model. Random or directed stimulus can be applied to the implementation and the software algorithm in parallel, and responses can be compared. The quality of checking is determined by the fidelity of the original software algorithm to the hardware implementation.

If the model is cycle-accurate to the implementation, detailed checking can be done. If the model is untimed or behavioral, the checking can be more difficult and less accurate. If the algorithm is implemented by running the software algorithm on a standard processor, the focus of the verification is on integrating the processor into the system. This verification is similar to the hardware platform discussed earlier. Verification is accomplished with the use of a BFM or ISS to speed simulation and provide control and visibility. Final verification should be done with the implementation model of the processor running the software algorithm.

ABSTRACTION

When you abstract information from an object, you take only the information that is relevant to your purpose and remove the rest of the details. Removing the details that are not important allows you to represent and analyze larger and more complex amounts of information. Abstracting information does not make that information less accurate. Many people believe that the more abstract a piece of information is the less accurate it is. Accuracy measures the correctness of the information. Information as it is abstracted must remain accurate. Fidelity measures how closely the information represents the original details. The fidelity of the information decreases as information is abstracted, but the accuracy must not.

Design Abstraction

Abstraction has made the design of complex electronic systems possible and can be beneficial in the verification of these same systems. At the most basic level, an electronic design is simply the flow of electrons through different physical materials. The most basic representation of a electronic system is the description of the physical materials, along with the charges applied to the materials. This level of detail was sufficient for describing very small primitive electronic behaviors. As designers wanted to represent more complex circuits, they needed to abstract the information from the circuit that was most relevant to the design. In this case, designers abstracted from the physical layout details the functional components that these details represented. The designer could then design using components such as transistors, resistors, and diodes. The facilitator for moving to this higher level of abstraction was the development of circuit simulators, such as SPICE, along with the associated libraries and netlisting facilities. These automation tools allowed designers to represent and analyze their designs at the component level, often referred to as the transistor level.

As designs became larger and more complex, designers again needed to move to a higher level of abstraction to more efficiently represent and analyze their designs. They abstracted from the component details of transistors and diodes to the Boolean gate level. This allowed the designers to simply specify the Boolean gate types that are created from various components. The facilitators for this were schematic capture, Boolean optimization, and analysis tools. Once again this level of abstraction worked well until the size and complexity of the designs outgrew the effectiveness of this level.

The next level of abstraction was not as easy to define as the standard component or Boolean gate level. The next level needed to represent the behavioral characteristics of the design in a more efficient manner, but there was not an established standard for this representation. For several years different design teams used different behavioral descriptions for this next abstraction level. These different and incompatible levels of reference made it difficult for tool development to facilitate a large move to one level. The facilitator became logic synthesis based on an industry standard description language called Verilog. This resulting level abstracted the functional representation of synchronous designs from the logic gates by specifying the functional operation between clock cycles or registers. This level became known as RTL. Most designs today are written at RTL, which is then translated to the logic gate level by logic synthesis, which maps to gate libraries specifying component-level detail, and finally to the layout level of detail.

Many attempts have been made to again raise the level of abstraction. Some success has been made in representing designs at an algorithmic or behavioral level, but the loss of detail in moving to this level has resulted in less optimized designs. At some point, the demand for larger and more complex systems, along with design automation tool breakthroughs, will move the design representation once again to a higher level of abstraction. Most experts today feel that the physical design and functional verification of today's designs must improve before design size and complexity grows to the breaking point.

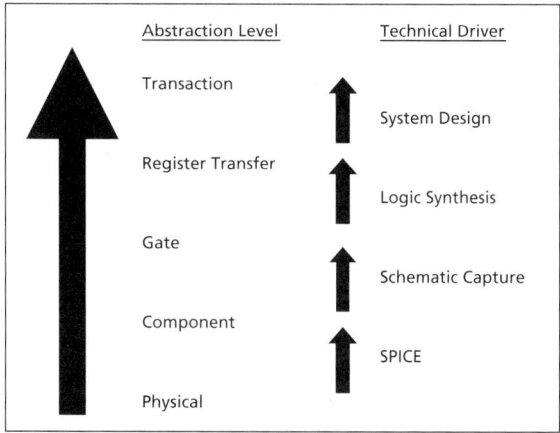

Figure 44. Design Abstraction Levels and Technical Drivers

Verification Abstraction

Verifying electronic systems has always followed the lead of design in representing and abstracting information. At the physical or transistor level, the generation of stimulus is quite simple in spanning the operating region. The fabrication process, temperature, and frequency are the important simulation parameters to test over the operating region. Analysis at this level is focused more on AC and transient effects than on functional behavior. At the gate level, Boolean verification could often be done by brute force applying test vectors to cover all possible test cases. Once designers moved to RTL, there was a need for more complex testbenches with stimulus generation and response checking. These testbenches were written in the same register transfer language used for system design. As RTL designs became more complex, RTL became more cumbersome for developing complex testbenches. The need for improved verification efficiency was first addressed with specific verification languages, which added optimizations for coding and data representation to RTL. These languages suffered in performance, because they were still tied to RTL and were fragmented due to the inability to standardize on one language.

Advanced verification teams have come to realize that verifying today's large complex systems requires moving to a higher level of abstraction. They cannot wait for designs to lead the way to the next abstraction level. RTL is sufficient for verifying low-level details, such as signaling of protocols and handshakes at interfaces, but it is not sufficient for representing the algorithms, data structures, and data flows within a complex system. Advanced verification teams have begun to move to the transaction level of abstraction, which removes the low-level signaling details and represents the algorithms and data flows in a more efficient manner. Removing this level of detail makes definition easier, simulations faster, and allows teams to begin before all the low-level details have been defined. The transaction level still retains the implementation-specific information necessary for verification and facilitates the refinement of this information as the design develops. Moving to an even higher level, such as a pure behavioral level of abstraction, would remove necessary information for verification.

Currently, the definition of the transaction level suffers from the same lack of clarity that RTL did before logic synthesis was introduced. Many teams are working at the transaction level, but they each have slightly different definitions of what level of detail is needed. Some teams specify the transaction level to the cycle boundary, where the information at each clock cycle is accurate to RTL. This cycle-accurate definition is only slightly higher than RTL and might be the next step for design, but in most cases, it is still too low a level for verification. Other teams only specify the behavior of the design and include only the most basic forms of timing and synchronization information. The driver for RTL was the move to a common logic synthesis tool—teams understood that it was in their best interest to converge on a single accepted representation. The driver for the definition of the transaction level will be driven by verification and more specifically the testbench. The industry will converge on a single representation of the transaction level to facilitate the development, transfer, and reuse of models and testbench components, because it is for the common good of all. Teams today have settled on definitions of the transaction level. This definition has allowed these teams to define interfaces between testbenches, models, and tools. The infrastructure for verification is built around this definition. As the different definitions converge into one, the industry will be able to leverage this infrastructure and optimize efficiency.

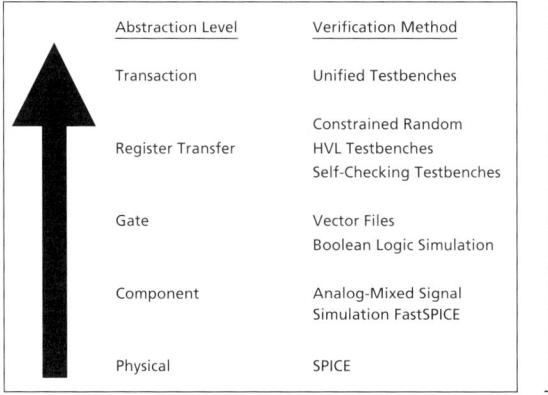

Abstraction Level	Verification Method
Transaction	Unified Testbenches
Register Transfer	Constrained Random HVL Testbenches Self-Checking Testbenches
Gate	Vector Files Boolean Logic Simulation
Component	Analog-Mixed Signal Simulation FastSPICE
Physical	SPICE

Figure 45. Verification Abstraction Levels and Associated Methods

Transaction-Level Modeling

The reason we have included this discussion of abstraction is that the system model should be written at the transaction level of abstraction. The system model should be started early in the development process before all the implementation details are available. The transaction level allows the definition of the model with basic implementation information and the refinement of the model as information becomes defined. Users of the system model require faster simulation speeds than available in RT models and cannot wait for the RT model to be completed. The transaction-level system model can be developed before the implementation is started, and can be simulated much faster as the low-level details are removed. Finally, developing the system model at the transaction level facilitates its use in the verification environment. Using a standard transaction-level definition, the testbench can interface directly to the system model to do early architectural testing and facilitate developing and testing of the verification environment.

A TLM abstracts the functional and data flow information from the design, removing the low-level signaling information. The functional or algorithmic information in the design is represented in the simplest manner. The model does not care about the implementation specifics within the function. The model focuses on the correct functional operation and the interface between different functions. A TLM uses a common representation of transfer of information between functions. This representation of data, and the transfer between functions or blocks, is often defined as a transaction. A transaction abstracts all the low-level signaling and handshaking information in the exchange of information between functions in a design and represents the data in a data structure with associated transfer information, such as the number of clock cycles the transfer will take. Thus, a TLM is made up of behaviorally defined functions interfaced together using a common transaction interface.

One way to think of the TLM is as a high-level function wrapped in an interface layer, which provides the transaction interface. As the function is decomposed down to smaller functions for implementation refinement, each sub-function is again wrapped within a transaction interface. Using the transaction interface facilitates easy replacement of functions, easy interface to analysis tools, and easy interface to verification or testbench components. A TLM can be written in an RT design language, but remember that the infrastructure and interface to that language will still be tied to RTL, so performance may suffer. The easiest language for defining functions as well as architecture analysis and software interfacing is usually a language based on C or an object-oriented language like C++. The industry has developed modeling and verification extensions to the C++ language called SystemC. Many companies are developing a common transaction level infrastructure using the SystemC extensions. The process and semantics of writing a transaction level system model in SystemC are quite complex. There are several excellent texts that focus on the mechanics of creating a SystemC TLM.

Chapter 13

Formal Verification Tools
Understanding their strengths and limitations

There has always been a great deal of confusion when discussing the topic of formal functional verification. Formal verification uses mathematical techniques to prove that a design is functionally correct. These mathematical techniques are steeped in theory and mathematical sciences that go beyond the comprehension of most verification engineers. The complexity of the underlying technology leads to much of the confusion when discussing formal verification. Rather than focus on the underlying technology, in this chapter we will look at using formal verification tools to address functional verification issues that advanced teams face today. We will also examine the limitations of these tools and techniques. As with all techniques, it is important for each verification team to measure the time and effort involved against the return received as well as whether the strengths outweigh the limitations.

WHEN TO USE FORMAL VERIFICATION

The original intent of formal verification was to address simulation's inability to cover all the possible test cases or state space in a design. This is still a large issue today, and formal verification is continuing to attempt to address it. But until this issue is addressed, the best use of formal verification is for smaller, more practical issues, such as locating basic bugs.

Formal verification can find the easy implementation bugs quickly so more time and effort can be applied to more difficult verification tasks. Static analysis tools, also called linting tools, use formal techniques to identify basic bugs without requiring a testbench or test stimulus to be developed.

Formal verification is also used to reverify a design when changes occur. Synthesis tools are commonly used in the development process to synthesize a gate-level netlist from an RTL representation. Assuring that this automated process worked correctly is important, since the verification typically occurs only on the RTL. The design often changes due to implementation changes, such as adding a clock tree or scan test logic. Some teams attempt to repeat the verification at the gate level, but the decrease in abstraction level results in performance degradation. Repeating the entire test suite at the gate level might not be possible in the time available.

To address this, formal verification uses mathematical techniques to compare the original representation to the new representation. These tools, known as equivalency checkers, break the design down into mathematical representation and then formally prove that the two are equivalent. Equivalency checkers can verify that two representations of a design, such as RTL and gate level or gate level and transistor level, are functionally equivalent. So once the verification is done at one level, it does not have to be repeated at the other level. You can also use equivalency checking to verify that the change made to a design only affects the intended functionality and does not have other functional implications.

Difficult bugs are often not found with traditional simulation approaches. Large complex designs have a huge state space, which can be impossible to cover with simulation. Teams might employ random stimulus simulation techniques to cover as much of the state space as possible, but these techniques are random and tend to take the path of least resistance. If a bug exists outside of the covered state space, it will most likely not be found until the design is in silicon.

While formal verification tools might lack the capacity to cover the state space of a large design, it can thoroughly cover selected important areas that are difficult to test with simulation, such as an arbiter or a complex queuing scheme. Formal model checking tools cover these areas of the state space by focusing on only certain parts of the design. Semi-formal or hybrid formal verification tools use simulation to lead the tool to interesting places in the state space and then thoroughly verify around that area. These approaches allow verification teams to focus on the hard to simulate areas where difficult bugs are found.

FORMAL VERIFICATION TECHNIQUES

Advanced verification teams have learned the advantages and disadvantages of using formal verification tools by dedicating large amounts of time and effort to them. These teams have learned where the time and effort has paid off and where it has been wasted.

RTL Analysis (Linting)

Design and verification teams have used lint tools for many years. RTL analysis tools take the design RTL as input and analyze the code for bugs, without a testbench or specification of properties. These tools use a variety of methods for performing this analysis, including formal techniques.

Figure 46. RTL Analysis Tools

The major deficiency in most RTL analysis tools is the trade-off that must be made between accuracy of the check and the chance of incorrect errors being reported. These tools cannot infer all the information necessary to absolutely guarantee that the issues found are real bugs. If the tools follow the strict coding of the check, they might report many issues to the designer that turn out to be correct design characteristics. If the tools try to limit the checking or infer more information than is really available, they risk missing a bug and giving the user a false sense of security. It is this trade-off that has moved RTL analysis tools to more complex methods to find real bugs with fewer false errors.

The simplest form of RTL analysis tools parse the code and identify issues from the textual representation of the design. These tools can find typographic and syntax errors that might result in incorrect connections or missing logic. Unfortunately, simply parsing the text does not provide enough information to detect more complex bugs without the risk of reporting many false violations. Early tools that focused only on the text were infamous for reporting thousands of violations that were not real errors. To overcome these false violations, RTL analysis tools began to translate or elaborate the design into a Boolean or mathematical format so that more complex information could be inferred. This information can help identify cases that can be logically proven to be correct or incorrect. An example of this is a net being driven by a tri-state device. By breaking that circuit down into its Boolean-level representation, the tool might be able to identify that it is logically impossible for the two drivers to be on at the same time, thus it would not report that as a false violation.

Elaborating the design can identify some information to provide more checks and limit the amount of false violations, but it still does not provide a complete solution. Many checks require a more complex mathematical analysis to prove the design meets the expected behavior. The latest generation of tools uses formal techniques to prove that certain complex bugs are real. These checks verify complex relations, such as deadlocks, clock domains crossings, and reachability.

Figure 47 shows a simple circuit with two different clock domains. A possible race condition occurs between register A and register B because it lacks the synchronization registers that exist in the path between register A and register E. Detecting this race condition is almost impossible using simulation, but an RTL analysis tool can easily identify this violation.

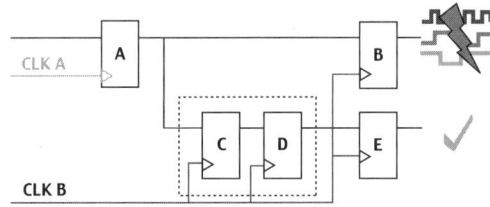

Figure 47. Clock Domain Crossing Bug Detection

Quite often the inability to determine whether a violation is correct is due to the inability to determine the behavior of design inputs. Tools might report that a violation will occur if a design input behaves in a particular manner. The designer looks at the violation and realizes that the input will never behave in that manner, so the violation is false. The only way these false violations can be removed is by providing more information on the correct behavior of the inputs. Many of the newer tools let the user specify properties or assertions at the inputs of the design. These assertions constrain the tools to only identify violations that occur without violating the properties or assertions specified at the inputs. Providing this information requires additional work, but offers the benefit of analyzing fewer false violations. This still does not guarantee that the tools will find all bugs. In many cases, the logic is too complicated due to size or state space, so the tool cannot prove that the check is correct. In these situations, the tool notifies the user that the check was inconclusive and that it should be promoted to simulation.

Most advanced verification teams use an RTL analysis tool, but differ on how they use it and how much they rely on it as a part of their methodology. The trade-off usually breaks down between the amount of time and effort spent on running and analyzing the results and the amount of bugs or design errors found. Different teams have different thresholds for time spent with an analysis tool versus simulation. This threshold often reflects the discipline or structure of the team and its management. Teams that are very regimented with disciplined and metric-oriented managers tend to use RTL analysis more than more open and relaxed teams.

There are three basic use models for RTL analysis tools:

- The designer or verification engineer takes each block of code as it is written and runs it through the analysis tool before simulation begins.

This removes easy bugs and assures quality code before it enters the verification process.

■ Teams run the tools every time a design change is made to verify that the changes have not caused a new violation.

■ Teams run the tools as a final verification before code is signed off. This may be part of a formal code review process.

■ Advanced verification teams have learned to use RTL analysis tools in the context of their entire methodology. The following lessons learned from these groups can be valuable to any team using RTL analysis tools:

■ Understand what you want to achieve and know the types of bugs that you want to find and that are important to you. If you do not do this analysis first, you might waste lots of time investigating violations that are of little value to you.

■ Understand what the checks are really verifying. Quite often two tools use the same name for a check, such as clock domain crossing checks, but what they check is very different. You can fool yourself into thinking that you have verified an aspect of your design that is not fully checked.

■ Set up defined rules and processes for using the tools, such as when is the tool run and what results are considered acceptable. Many teams add their own checks to the tools to verify their own design style requirements.

Equivalency Checking

Equivalency checking is not always thought of as a functional verification technique, because it is mostly used during the implementation or back-end stage of the development process. Equivalency checking compares two representations of a design to verify that they are functionally correct. Functional equivalency means that the two representations are logically equivalent. If you apply the same stimulus to the two designs, you will get equivalent functional responses. Equivalency checkers break the design representations down to a basic mathematical representation and then use formal techniques to prove that the representations are equivalent.

The original promise of equivalency checkers was that they could provide the verification link between high-level behavioral representations developed by architects and the low-level implementation representations developed by designers and automated tools. Unfortunately, the higher in abstraction a design is modeled, the more ambiguous the representation can be. Verifying the functional equivalency of two ambiguous models or an ambiguous model and a detailed model is very difficult. Thus, the promise of verifying an architectural or behavioral model to the final implementation is still unrealized.

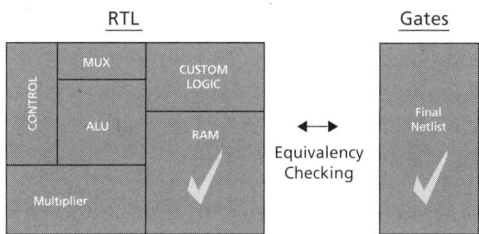

Figure 48. Equivalency Checking Between RTL and Gates

Verifying the equivalency of models at high levels of abstraction might be difficult today, but verifying the equivalency of more detailed implementation-level representations is possible. Equivalency checkers are being used today to verify the equivalency of designs represented at synthesizable RTL, the Boolean gate level, and the transistor level. Equivalency checkers are used to verify that the gate-level representation produced by a logic synthesis tool is equivalent to the RTL fed into the tool. This verification proves that the synthesis tool performed correctly and eliminates the need for reverification of the gate-level representation. Quite often portions of a design, such as memories, data paths, or pin interfaces, are not developed from an RTL representation synthesized down to a standard gate-level representation. These portions are created by hand with custom design methods. To simulate these portions, a model is created at a higher level of abstraction, so performance is not degraded. Equivalency checkers can be used to verify that the functional model of the custom-designed portion used during functional verification is equivalent to the design that will be implemented. Using equivalency checkers in these ways enables you to functionally verify the design once at a high level of abstraction and then prove equivalency to the lower implementation levels rather than reverifying the design at each level.

When two designs are not equivalent, equivalency checkers can also provide you with the exact functional differences. This information is useful when small changes are made, such as resizing components, adding test and clocking logic, and correcting minor functional errors. Instead of reverifying the entire design each time a change is made, you can use equivalency checkers to identify the exact difference between the original verified design and the new modified design. This comparison assures the development team that the change did not inadvertently affect some other part of the design that could cause a functional error.

The users of functional equivalency tools are most often the engineers responsible for logic synthesis, model creation, and final implementation. Most advanced verification teams are not involved in these functions, so their use of equivalency checkers is limited. However, verification engineers need to understand their use to be able to identify when they should be used and when reverification is necessary.

Model Checkers

Model checking tools are perhaps the most commonly associated with formal verification techniques today. Model checkers prove that a property or assertion about the design is true or false. Advanced verification teams use these tools in two different ways: to verify specific parts of the design that are most amenable to formal proofs or to find difficult bugs that the user had not thought of or was not able to simulate. Each of these uses can enhance your verification methodology.

Because of capacity and performance issues, formal model checkers have never been able to keep pace with the growing size and complexity of digital designs. Thus, the use of model checkers has been focused on parts of the design that are small enough to fit in the tools and are amenable to formal techniques. The strategy many advanced teams use is to identify areas that will be very difficult to verify with simulation because of the number of combinations of interactions that would need to be tested. Examples of these areas include complex control logic found around memory controllers and bus interfaces, arbiters, and complex queuing and flow control logic, such as leaky bucket or token-based algorithms.

Once the areas are identified, these blocks are given to a verification specialist who is familiar with model checking. Model checkers are very complex to run and debug, so a specialist is usually required. The specialist meets with the block designer to identify properties or assertions that need to be proven. The designer might be able to provide some assertions in the form that was embedded into the code, but often they need to be modified for the tool. The specialist also needs to understand how stimulus is applied to the design, and what legal or illegal stimulus is. Constraining the tool is perhaps the most difficult part. If the inputs or the assertions are underconstrained, the tool reports many invalid violations. If they are overconstrained, bugs could be missed.

The specialist runs the tool and identifies which violations are caused by incorrect constraints and which may be real bugs. The specialist identifies the bugs to the designer, who then determines whether they are real bugs or cases where more constraining is required. The specialist modifies the constraints or obtains a bug fix and reruns the tool to repeat the process. When the tool reports that no violations can be found, the design is considered verified, although many teams go back through the constraints to verify that the tool has not been overconstrained.

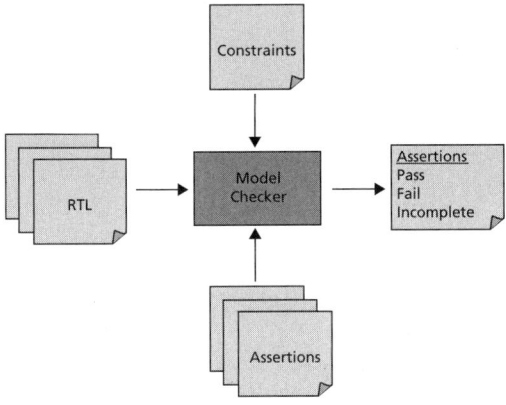

Figure 49.   Model Checkers

Model checking is a powerful way to focus verification on one area of a design, but it does have its drawbacks. First, the technology requires a specialist to operate. Many large companies can afford to hire one or a team of model checking specialists, but smaller or resource-limited companies usually cannot. Secondly, the design areas where model checking is useful is limited. Finally, model checking does not always come up with a complete proof of an assertion. It is common for the tool to report that it was unable to prove if the assertion was valid or not. If this happens, all the work to run the tool was of no value.

Semi-Formal Verification

Model checkers are also used in a newer form of formal verification called semi-formal verification. Semi-formal verification tools overcome many of the deficiencies of formal tools by combining formal techniques with standard simulation techniques. The premise is to find difficult bugs within your design that cannot be found with traditional approaches. These bugs are difficult to find with directed tests because they are so complex and obscure that you cannot possibly think of every possible scenario. They also cannot be found with random techniques, because the stimulus to trigger them is very complex or the number of possible scenarios is so great that the odds of randomly finding them are slim.

The original promise of semi-formal tools was that you could use simulation techniques to lead you to interesting areas in the design where bugs might be hidden and then use formal techniques to expand from that location to formally prove all the possible combinations of events around that area as shown in Figure 50.

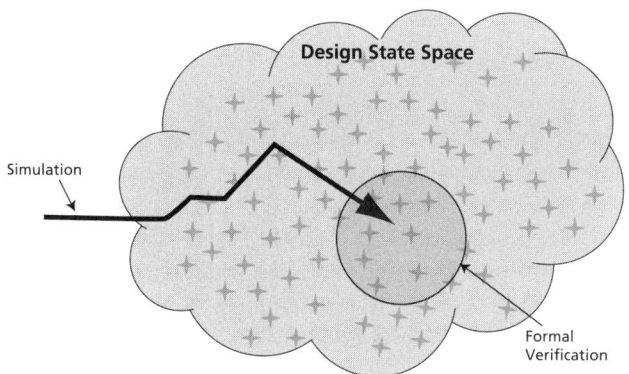

Figure 50. Simulation Finds Interesting State, Formal Proves Area Around

The promise of semi-formal tools has not yet been realized because of the difficulty of automatically finding interesting states to start from and detecting bugs once you do find the area. Automatically identifying interesting areas requires a target for the tool to search for. The most common method is to search for coverage metrics like state machine arcs or process interactions. Unfortunately, these do not always lead to the most interesting places and can often send the tool off into unimportant areas, such as test logic. A workaround is for the user to supply information to the tool. Some tools use an already created directed test as a jumping off place to engage formal verification techniques by formally verifying the design starting from the existing states of the test. Another group of tools uses assertions placed in the design to target the formal verification engines. These tools simulate the design until the assertion is stimulated, and then they engage the formal engine to verify possible scenarios starting from that point. Directed tests or assertions help direct the semi-formal tool to interesting areas, but these areas must first be identified by the user. If a difficult bug is not found in an identified area, the odds are slim that it will be found with these tools.

The other issue is whether semi-formal tools have the ability to check for bugs once an interesting area has been identified. The most common form of checking used with testbenches is data checking. Data is applied to the design, and the results are compared with expected results at the output of the design. This is true for a design as complex as a router or as simple as an FIFO. Data checks require the ability to track or store data as it moves through a design and the ability to predict the correct response. Formal verification techniques require that the design as well as the verification logic be understood by the tool. Thus, formal techniques are targeted more at sequence or protocol checks that verify the relationship between signals or events. Formal verification cannot handle large arrays of data storage required for data tracking and cannot understand data prediction and checking operations required for data checks. These limitations result in data checkers being removed from the formal proof. The only checks made while the semi-formal tool is looking for difficult bugs are the assertions placed in the design. If the tool does find an interesting state and covers a scenario that can cause a bug, that bug is only detected if one of the assertions catches it.

Even with these limitations, today's semi-formal tools still provide another mechanism for increasing the likelihood of catching a difficult bug. These tools can handle larger and more complex designs and are easier to use than traditional model checkers. Advanced verification teams use these tools as an additional mechanism to find bugs that they might not have been able to find before. Unfortunately, the limitations and the inability to provide coverage information means that these tools do not replace existing tasks but rather they add to your already complex verification task.

Chapter 14

Testbench Development
Measuring the trade-offs

Simulation is by far the most prevalent technique used in functional verification today. The ability to verify the ideas as well as the implementation before a device is manufactured saves a development team time and effort. Developing a testbench environment is often the single most important and time-consuming task for an advanced verification team. Many excellent classes and texts on how to build a testbench for various types of designs using various languages and tools are available. This chapter presents some important issues and trade-offs that verification teams need to consider before building a testbench. It also describes how to develop a reusable unified advanced testbench.

Teams should focus on three basic goals when developing a testbench: efficiency, resusability, and flexibility. The testbench should make verification more efficient by removing the low-level details and redundant processes so that the verification engineer can focus on testing and debug. The testbench should be designed to facilitate reuse of its components within other similar testbenches. The testbench needs to flexible so that it can be easily leveraged and integrated with other environments. It should facilitate the integration of different designs and support the integration of the design being verified. These three goals often conflict, forcing the testbench developer to make trade-offs to create a testbench most suitable for the intended use.

TRADE-OFFS

Testbench developers have been striving to meet the goals of efficiency, reuse, and flexibility for many years. Unfortunately, attaining these goals often makes testbenches more complex to create and more difficult to use. Every testbench developer must make a trade-off between the time and effort to create and use the testbench versus the potential gain from making the testbench efficient, reusable, and flexible.

Reuse—Isolating Design-Specific Information

To improve the reusability of a testbench, a developer should focus on isolating the design-specific information in the testbench and separating the functionality of the testbench. Whenever information that is specific to the design being tested is embedded in parts of the testbench, it becomes less reusable. The application-specific information most likely will need to be modified before the testbench can be reused for a different design.

A testbench provides several basic functions, including creating and applying stimulus and verifying the correct interfacing and responses. Each of these functions can have design-specific properties, such as the stimulus generator generating a certain type of data and the device interface applying data to a specific type of bus. Separating these functions into different components allows the functions to be reused in a different configuration for a different testbench.

Partitioning and separating information takes time and effort. Knowing about the number of cycles after stimulus the response appears, or observing internal design signals to help predict the expected result, can simplify testbench creation. However, once this information is embedded in the testbench, it is difficult to reuse the testbench for a different design or modification of the design. Partitioning the design into different pieces also comes with the overhead of defining more interfaces and maintaining more modules.

Most advanced verification teams find that small internal blocks with non-standard interfaces are the worst candidates for reuse. Testbench reuse is most advantageous at the subsystem level, where interfaces are more standard and the testbench components more complex.

Efficiency—Abstracting Design Information

To improve the efficiency of a testbench, a developer should abstract design information to a higher level. The testbench should represent data and actions in a format most easily understood by those using the testbench. Test writers should be able to write their tests at a functional level consistent with the application. Low-level implementation details that are irrelevant to the test should not be specified. Throughout the testbench, data should be captured and compared at a higher level of abstraction to make debug easier.

The most important step in abstracting design information is choosing the correct higher levels to abstract to. The levels chosen should be consistent across the testbench so that common test infrastructure and analysis mechanisms can be used. Once the abstraction levels are chosen, converters can be created to facilitate the abstraction. A converter can be created to convert a high-level test language used by a test writer to the stimulus generator tasked with creating the specific stimulus. Converters can also be created at interfaces throughout the testbench to convert from the low-level implementation details to the higher abstraction levels.

Abstracting design information in a testbench requires extra work and produces extra testbench code. Developing converters is not a trivial task and always has the potential for introducing new bugs. The gain in efficiency obtained from abstraction often comes at the expense of controllability and observability. Removing the implementation details from the test writing process limits the control the test writer has to affect the stimulus. Abstracting implementation details from design data for checking and debug might also overlook important implementation characteristics, and possibly bugs.

Most advanced verification teams find abstracting design information to be valuable for newer designs that require a large number of tests and long debug cycles. Existing designs with existing stimulus sets do not benefit from large efficiency gains from abstraction.

Flexibility—Using Standard Interfaces

To improve the flexibility of a testbench, the developers should focus on utilizing standard interfaces to facilitate multiple different uses. The developer should use standard interfaces for all tools and processes associated with the testbench. Examples of these interfaces include testbench development languages, simulator interfaces, and debug methods. Developers should not be trapped into using only a limited set of options for tools and processes associated with the testbench. They should focus on industry standards supported by two or more independent parties and controlled by an independent organization, like the IEEE.

The first step in utilizing standard interfaces in a testbench is clearly segmenting the intended functions of the testbench from the function of external tools or processes. Developers should focus their energies on the functions of the testbench that provide the most unique value. Once this segmentation is determined, the developers should standardize on an interface to the testbench that allows for flexibility and adaptability in connecting to the testbench.

Using standard interfaces might rule out using some of the most leading edge tools and technologies for developing a testbench. Any standardization process takes time, and developers might not want to wait for multiple parties to agree on a solution. In these cases, developers should make sure that their interfaces to new tools and processes can be changed in the future if a better interface comes along so that they do not get locked into a proprietary interface.

Advanced verification teams develop their testbenches independent of the tools or languages used. The environment of the language used should not dictate the architecture of the testbench. These teams make sure their testbench is adaptable so that they can easily switch tools or technologies without changing the testbench architecture.

Balancing Practical Concerns

Developing a testbench is a balance of many different factors. Knowing the scope of the verification task helps determine the size of the testbench and the requirements for reuse and integration. Is the goal to verify an entire system from the ground up or to verify an individual device or block within the system? Do you need to verify the integration of software, analog, or purchased IP within the context of the testbench? Verifying a complete system from the ground up requires a testbench that can verify from the small block level up to an integrated subsystem and to a final system level. Verifying a system made up of existing parts might just entail developing a testbench that verifies the interconnection and interoperability of the different parts. Verification of a single device or block might limit the necessity for reuse within the testbench.

You should consider the type of design being verified. A design where data flows through the design from input to output with some translation occurring to the design might require a basic single driver and checker testbench. A design that controls a central resource where data passes in multiple directions with different input and output ports might require a more centralized testbench. In general, a testbench should be a reflection of the environment that the device will operate in. If that environment is similar to a data path, the testbench should operate like a data path. If the design is a central controller, the testbench should operate like a group of centrally linked resources.

The testbench developer also needs to know how thoroughly the design needs to be tested within the testbench environment. If the testbench is the only place the design will be tested before its manufacture, the testbench needs to simulate all possible corner cases, including illegal operations. If the design will be integrated and tested with a different testbench or prototype, the testbench developer might be able to be less thorough.

Another factor to consider is what test and design information is available. If a system model is available, the testbench developer might be able to use this within the response checker or as a substitute for a design block. If stimulus files with the correct response information are already available, the testbench developer can use these in place of stimulus generators and response checking. If standard algorithms or specifications that can be easily converted to a model are available, the testbench developer might be able to utilize this information.

You should also consider the practical aspects of the verification task. The program schedule will most likely dictate how much time the developer has to create the testbench. The number of resources and the skill level of the developers should be factored into how the testbench will be implemented and how complex it will be. The number of test writers and their skill level should also be factored into the interface to the testbench.

Top-Down vs. Bottom-Up Testbench Development

When developing a complex testbench to verify a device at multiple levels of hierarchy, the most basic question to ask is where to start. There are two schools of thought when developing testbenches in a hierarchical manner. Some teams believe that it makes most sense to start at the lowest level of hierarchy and develop the testbench to first verify the units or blocks that make up the design. As the units are integrated into blocks and the blocks are integrated into subsystems or systems, the testbench is developed in the same bottom-up approach. Other teams believe it is best to develop a testbench from the highest level of the hierarchy and derive the testbenches for each lower level from the higher levels. As the system is partitioned down into subsystems, blocks, and units, the testbenches are partitioned in a similar top-down manner.

Whether to develop a testbench in a top-down or a bottom-up method depends on several factors. First, is a system model available. The top-down approach is usually most applicable to teams that are using a system model in their verification environment. The bottom-up approach is usually used by teams that only have a written specification. A second factor is the necessity to integrate other design domains and software development into the verification process. The top-down approach is beneficial when the design requires integrating software or analog/RF domains at the system level. Thirdly, what is the availability of resources? The top-down approach offers greater benefits if the development team has a separate verification team to write tests and develop the testbench in parallel with the implementation of the design. If the development team relies on the designers to perform verification or the verification team is not engaged until after the design is complete, a bottom-up approach may be more efficient. Then there is reuse. The top-down approach is able to more fully utilize reusable testbench components. Finally, what is the skill level of the verification team. A bottom-up approach requires less knowledge of complex verification, such as system model development and reuse. A testbench developed from a top-down system model requires more maintenance to keep the model and testbench up-to-date with the design.

Because of these benefits and requirements, some development teams choose a more specification-based bottom-up approach to subsystem verification for today's designs. However, as the use of processor-based SoCs increases, along with the further integration of digital and analog on the same chip, the benefits of a top-down methodology will cause many teams to migrate. In Chapter 8, we presented a top-down flow for SoC-based designs as a standard for teams to migrate toward. We also presented a bottom-up specification-based flow that utilizes many of the advanced techniques of the top-down flow as a place to begin the migration.

There are many similarities between these two flows. Even though the bottom-up flow develops the testbench from the lowest unit level up to the subsystem level, the planning and architecture for the flow is top-down. The only way to create testbenches that can be reused as the design develops from the unit to the block to the subsystem level is to plan ahead and know what will be required at higher subsystem levels. Similarly, even though the top-down flow develops the testbench from the top SoC level down to the lowest unit level, testing and integration are still performed in a bottom-up manner. Both flows use transaction-based testbenches for performance, assertions for easy debugging, coverage for efficient test development, and hardware acceleration for increased performance.

The two flows differ in the development of the verification environment and tests. In the top-down flow, the environment is developed from the highest levels down to lower levels using common models and testbench components. The SoC-level environment is developed first with a system model, the system model is refined down to the block and sub-block level, and the testbench is developed in the same manner. This enables the verification engineers to use the model to test their code, allows reuse of the models in the system model at different levels, and promotes parallelism and accelerated integration of the implementation when it is made available.

The bottom-up flow develops the testbench from the lowest level up to higher levels. The flow reuses testbench components from the lower levels as the design is integrated and tested. Since a common model is not used, reference checkers might need to be developed at each level or linked together. Also, the lack of an accurate model means the tests and testbench components are developed in isolation until the implementation is available. This might cause the simultaneous debugging of implementation and testbench.

UNIFIED TESTBENCHES

Chapter 8 discussed verifying digital subsystems within the context of the UVM. The development of a unified testbench is vital to attaining the UVM goals of increased speed and efficiency. Here, we describe a high-speed, reusable testbench that meets the requirements of the UVM.

Testbench Components

High-performance, reusable testbenches are based on standard components with a common interface for communications at different levels of abstraction. Figure 51 shows the basic components.

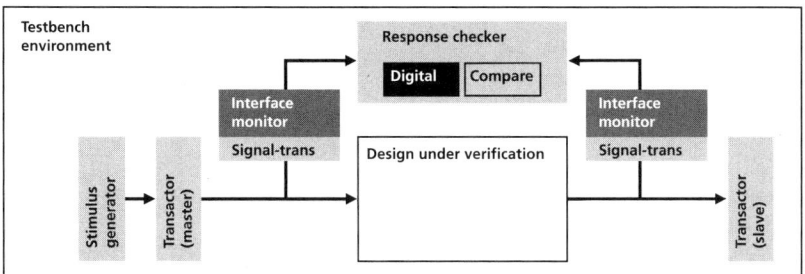

Figure 51. Unified Testbench Structure

Stimulus Generators

Stimulus generators create the data the testbench uses to stimulate the design. Stimulus generators can create the data in a preprocessing mode with custom scripts or capture programs, or they can create the data on-the-fly as the simulation occurs. Stimulus generators are usually classified by the control the test writer exerts on the generation of the stimulus.

Transactors

Transactors change the levels of abstraction in a testbench. The most common use is to translate from implementation-level signaling to a higher level transaction representation or the reverse. Transactors are placed in a testbench at the interfaces of the design, providing a transaction-level interface to the stimulus generators and the response checkers. Transactors can behave as masters initiating activity with the design, as slaves responding to requests generated by the design, or as both a master and a slave. The design of a transactor should be application-independent to facilitate maximum reuse. Application-specific information can be contained in the stimulus generators or TLMs attached to the transactors. Also, when developing a transactor, the designer should consider its use in a hardware accelerator. Developing the signal-level interface in a synthesizable manner allows it to be accelerated along with the design, improving the performance gain obtained from a hardware accelerator.

Interface Monitors

Interface monitors check the correct signaling and protocol of data transfers across design interfaces. In some testbenches, interface monitors are combined either with the transactors or with the response checkers. Keeping interface monitors separate from these components allows for maximum reuse of the monitors. In addition to passively monitoring the data transfers across interfaces, these monitors can encapsulate the data to be communicated to response checkers. This allows the response checkers to concentrate solely on verifying correct operation. Interface monitors contain interface assertions and can be written in an assertion language. The interface monitors should be application-independent and written in a manner that allows their easy reuse in hardware acceleration.

Response Checkers

Response checkers verify that the data responses received from the design are correct. Response checkers contain the most application-specific information in the testbench and usually can only be reused when the block they are monitoring is being reused. There are three basic types of response checkers:

- Reference model response checkers apply the same stimulus the design receives to a model of the design and verify that the response is identical to the design. The most efficient method is to reuse the TLMs of the FVP for the reference models in the response checker.

- Scoreboard response checkers save the data as it is received by the design and monitor the translations made to the data as it passes through the design. The translations are tracked with a scoreboard, and responses are verified as they are generated by the design.

- Performance response checkers monitor the data flowing into and out of the design and verify that the correct functional responses are being maintained. These checkers verify characteristics of the responses rather than the details of the response.

Scoreboards are used when the design responds in a predictable manner and the data is easily correlated, such as a bridge or a switch design. Reference models are used when the design can be easily modeled independently with enough accuracy, such as a computation unit or a pipeline. Performance checkers are used when the functions in the design are unpredictable due to implementation specifics, and the correct operation can be specified by the characteristics of the design, such as a rate limiter or a routing algorithm.

Testbench API

The testbench API facilitates the communications between components in the verification environment at different levels of abstraction. The API is the glue that holds the testbench together. Using a standard API allows for the reuse of components within the verification process and between different projects. The testbench API defines the interface layers between components and between different abstraction levels.

Top-Down Testbench Development

Top-down testbench development starts with the development of a top-level subsystem testbench environment once the FVP and specification are delivered to the team. The subsystem testbench reuses some components from the FVP, as shown in Figure 52.

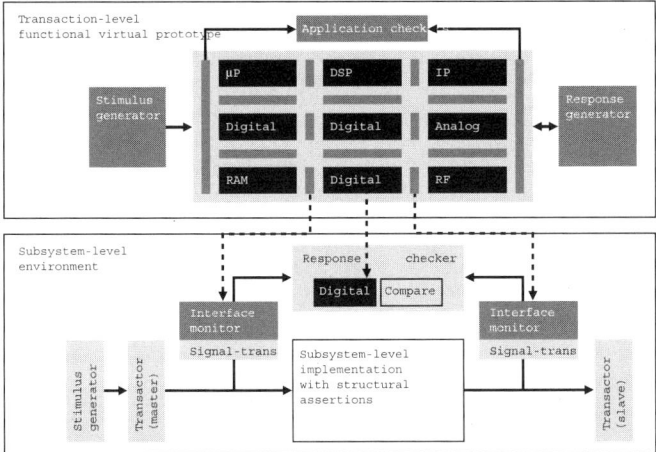

Figure 52. Reusing an FVP in the Testbench

The SoC team might provide signal-level transactors, or the subsystem team might develop them. The subsystem team develops the tests, which are verified in the testbench by substituting the FVP TLM for the implementation until it is available. This allows the test writers to develop and test their code before the implementation block is available, making the process more efficient. This is shown in Figure 53.

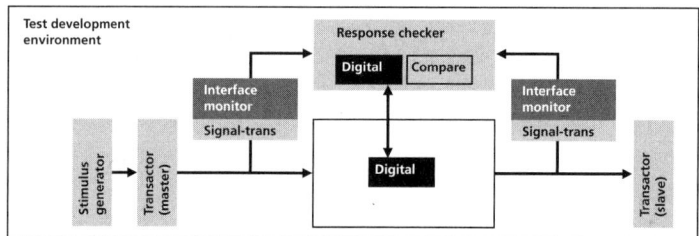

Figure 53.   Testbench Development Using a TLM

As the subsystem is micro-architected, it is partitioned into smaller hierarchical blocks and units. Large complex subsystems require the verification team to do much of the verification at the block level. The verification team chooses the levels of hierarchy to test at by selecting blocks with common standard interfaces that provide enough observability and controllability and are not too large for simulation tools to be efficient. Often designers want to verify at the lowest unit level as they develop individual modules. This is done by the designer using simple HDLs or HVLs, along with added structural assertions. The verification uses simple methods, applying stimulus vectors and waveform inspection or application-specific environments. These environments are not reused, since their purpose is simply to prove basic sanity of the unit-level modules.

Block-level testbenches are developed in a similar manner as the subsystem testbench. The FVP is partitioned to match the verification needs and reused in the individual response checkers. Transactors and interface monitors are reused from the subsystem testbench. Common transactors are used in slave or master modes for the different connecting blocks. The resulting testbenches are shown in Figure 54.

Tests can be developed in advance of the implementation in the block-level testbenches by substituting the models as described in the subsystem testbench. When developed in a correct and efficient manner, the top-down testbench method provides for the tests and testbench to be completed and debugged before the implementation is delivered by the design team. This creates the fastest and most efficient debugging and integration flow.

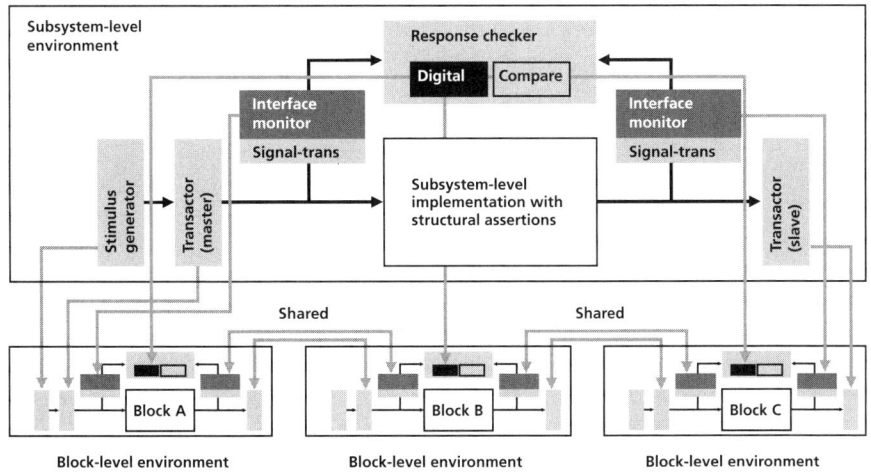

Figure 54.   Top-Down Reuse of Testbench Components

Bottom-Up Testbench Development

Bottom-up testbench development starts with the partitioning provided by the micro-architecture teams. When the partitioning is available, the verification team selects the block and subsystem levels to test at. Often designers want to verify at the lowest unit level as they develop individual modules. This is done by the designer using simple HDLs or HVLs, along with added structural assertions. The verification of these units uses simple methods applying stimulus vectors and waveform inspection or application-specific environments. These environments are not reused, because their purpose is simply to prove basic sanity of the unit-level modules.

Block-level development begins with building individual testbenches for the lowest block-level verification. New transactors, interface monitors, and response checkers are developed or reused from previous projects if available. The response generators can be based on models or modified from the FVP, but maintaining consistency between these models is difficult. Unless a behavioral model is available, the test teams must wait to run and debug their tests until the implementation is available.

The subsystem testbench is developed from the existing block-level testbenches, as shown in Figure 55. Transactors at internal interfaces are removed, and the response checkers are linked together with interface monitors to provide a complete subsystem response checker. Close communication must be maintained between block-level developers so that the subsystem integration works correctly. Blocks located closer to the stimulus must consider the effects of their environment on connecting blocks. Also, changes in the block-level testbenches must be propagated to the subsystem testbench and visa versa. Good communication and planning are required to make the integration of bottom-up testbenches efficient.

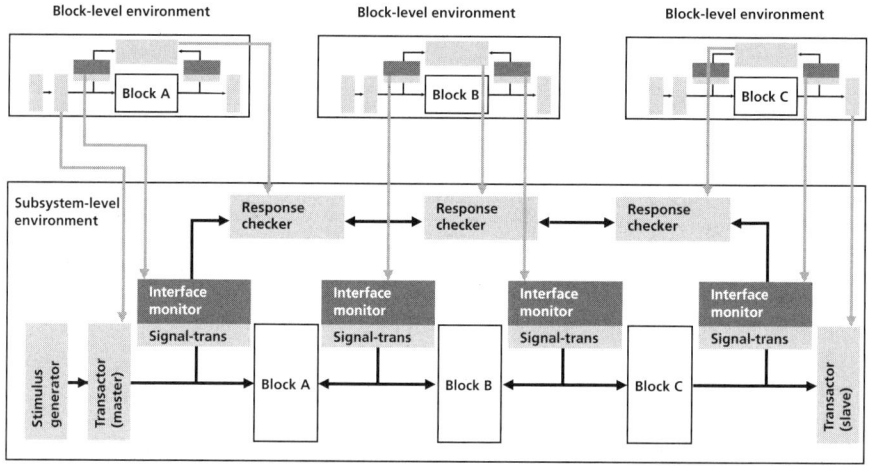

Figure 55. Bottom-Up Reuse of Testbench Components

VERIFICATION TESTS

So far we have only briefly talked about the actual tests that will run on the testbench being developed. Many verification teams separate the creation of the testbench from the creation of the test stimulus, because the two tasks are very different and require different skills. Usually a testbench is developed by a few engineers who are highly skilled at developing complex code and systems. Complex systems can require large teams of test writers that need to know the intricacies of the design and understand how to test that it is working correctly. Development teams often use design engineers and software engineers to write tests for a period of the development time.

The goal of the testbench developer is to create a testbench that allows the test writer to write and debug a test in the most natural form possible. The testbench should shield the test writer from the complexities of managing data and interfacing to the design. In this section, we will take a closer look at developing tests for an advanced verification testbench.

Directed and Random Tests

The two basic types of tests written today are directed and random. Directed tests specify the exact type and sequence of stimulus to provide to the design. Random tests automatically select part or all of the information for the test, including type, sequence, and timing of the stimulus. There are advantages and disadvantages to using random or directed tests. Most advanced verification teams use a combination of random and directed tests.

A directed test tests a specific function in a consistent, thorough, and predictable way. It is usually targeted directly at a specific feature or function with the stimulus coded specifically to stimulate and verify that operation. Using directed tests, you can incrementally test function after function and build up a thorough regression suite that can be used to reverify that function if the design changes. The disadvantages of directed tests are that they require detailed knowledge of the design being tested and are often very difficult to set up. Also, a large complex chip could require thousands of directed tests to fully verify the design. The time required to write these tests might not be feasible for the development schedule.

Random tests allow for the automatic creation of many cycles of stimulus with limited knowledge of the design required. You simply specify the range of values for the stimulus generator to generate data and let the test run for as long as you want. You only need to understand what stimulus is legal to apply and not the intricate implementation details of the design. One random test can verify many functions, so fewer random tests are required than directed tests. The disadvantage of random tests is that it is difficult to know what the random test has verified. You may need to insert monitors or use coverage information to understand which functions the random test stimulated and checked. Even if you can determine which functions have been verified, it is often difficult to consistently repeat the test for regression purposes. A complex random environment can behave completely different even if one small change is made to the test, testbench, or design. Thus, you might need to reverify what the random test had verified each time it is run. The other disadvantage of using a random environment is that it can be very difficult to debug. When debugging a directed test, the debugger knows what the test is designed to do. Because the random test determines the stimulus, it is difficult to debug a failure and to incrementally verify a design as the functionality is implemented.

Types of Directed Tests

There are several types of directed tests. Interface tests verify the correct operation of each of the major interfaces found in the subsystem. The tests verify correct handshaking and error handling. Stress tests are included to verify the constraints of the interface, such as time-outs, aborts, and deadlocks. These tests should be run first to verify the correct communication between the testbench and the design. Feature tests verify the processes contained within the subsystem. These tests verify the correct operation of the features under normal and stress conditions. Stress conditions include system interactions between features, such as interrupts, retries, and pipeline flushes. These tests are run second with the goal of verifying each feature in isolation under non-stress conditions before turning on randomness.

Error tests verify the correct operation of the subsystem to error conditions. Error conditions consist of recoverable and non-recoverable errors. Recoverable error tests verify the observation of the error and the recovery from the error. Non-recoverable error tests verify the observation and the correct system response, such as a halt, freeze, or interrupt signal. These tests should be run after feature testing is complete and random tests have run for multiple hours without failing. Performance tests verify that the subsystem meets the performance requirements of the system. Performance requirements can include latency, bandwidth, and throughput. The tests stimulate the system with normal rate stimulus as well as corner case stimulus that is known to be performance limiting. Performance tests are run periodically throughout the testing process to verify that the design is still meeting test goals and should be run again at the end of testing to verify that design changes and bug fixes have not violated the performance goals.

Combining Random and Directed Approaches

When random testing first became popular, many teams believed that the best approach was to first run random tests, measure what had been tested, and then write directed tests for the areas that were not stimulated by the random tests. In theory, this made sense; but in practice, it had many flaws. When a design is first verified, it can contain many basic bugs. The most efficient way to bring up a design is to test it in stages, applying very basic stimulus first to verify that it works before applying more complex stimulus. When using random tests, you often have limited ability to select in which order stimulus is applied, so a random test does not always follow the most efficient bring-up process.

It is also difficult to measure what the random test verified. Coverage tools can provide some information about what areas of the design have been stimulated, but they cannot tell whether a function has been verified completely. Some teams place monitors in the design that signal when a function has been tested. Placing these monitors throughout a design can be time-consuming and requires intimate knowledge of the design. Knowing at what point to stop running random tests and begin writing directed tests is also a challenge. The team needs to leave enough time for the directed tests to be written to cover the areas not stimulated by the random tests, but you do not know how many directed tests are needed until the random tests are run. Also, as noted earlier, the behavior of random tests can change with small changes in the design or testbench. This means that what is tested by random may change from run to run. Thus, it is often difficult to come to closure on exactly which directed tests need to be written. Often, a verification team thought they completed verification when it was discovered that a change in the testbench required another directed test to be written. This open loop process makes it very difficult for verification teams to know when the design is verified to a sufficient level to be taped out.

Today, advanced verification teams use a subset of directed tests first, then use random tests followed by more specialized directed tests. Advanced verification teams first develop a group of must-have directed tests that are used in bring-up and verify the most important functionality. These tests are also used as a consistent regression suite. After the first directed tests are run, the team can feel confident that the design is at a level of sanity where random testing will be of most use. The team can use monitors or assertions in conjunction with coverage tools to get a good idea of which functions have or have not been stimulated. At this point, a final set of directed tests are run to verify the special case conditions that were not hit by random. Having a suite of directed tests that verifies the most important functions eliminates the fear that random testing missed a basic function that pops up late in the development stage.

Constraining Random Tests

Random tests vary in the freedom of the randomness available when generating the stimulus. A pure random test is free to apply any stimulus at any time. This means that an input driven by the test could be set too high or too low at any time, in any sequence, with no restrictions. Pure random tests are rarely used because the lack of control makes the odds of the test stimulating an internal function very low. Instead, most tests constrain the randomness by limiting the data values, timing, or sequences to predefined ranges. By constraining the randomness, these tests can focus on stimulus that is of highest value and has the highest likelihood of stimulating internal functions.

The most basic constrained random tests randomize the data values across a specific value set. These tests randomly pick a value within a defined set of values specified by the test writer. For example, a test writer could specify that a CPU request be either a read or a write, access an address within a certain range, and wait between one and five cycles before starting the request.

More complex constrained random tests might specify weightings to the values in the defined set, making it more likely for the random generator to select one value or a range of values in the set over other values. For example, a test writer could specify that a CPU request be a read 10 percent of the time and a write 90 percent of the time, or that the operation accesses certain values within a defined range more often than other values.

As more constraints are added, it is more likely that the constraints overlap and form complex relationships. For example, a test might specify that a request be either a read or a write, but if the request is a read, the address range contains one set of values and if it is a write, the address range contains a different set of values. There is now a relationship between the request type and the address of the request generated. The random system must decide which random value to pick first. If the request type is generated first, the system must generate the address within the range for that request type. If the address is generated first, the system must make sure that the request type generated is the correct one for the address. Complex verification tests can contain hundreds of different constraints that all have relationships to each other. Verification systems require constraint solvers that use mathematical techniques to manage the correct selection of values to meet all the necessary constraints.

The more a random test is constrained, the more control the test writer has over the generation of stimulus and the more likely the test stimulates the intended functions. Also, the more a random test is constrained, the more similar it becomes to a directed test and the more design-specific information is required of the test writer. Test developers need to balance the need for controlling the randomness to target specific areas with the power of randomness to stimulate and discover areas that are unknown to the test writer.

Advanced verification teams utilize constrained random tests as a substitute for many of the directed tests they would need to write. One random test running for many cycles might stimulate the same functions as five or ten directed tests. Constrained random tests are also used to duplicate the random nature of traffic flowing into the system. Most electronic systems today operate in a very non-deterministic environment. Verification teams want to be able to verify that the system operates correctly in a similar non-deterministic manner. Constrained random systems can mimic the traffic patterns that might be seen by the device.

Testbench Requirements

Constrained random tests require a testbench that has been designed for supporting a constrained random approach. The testbench must be self-checking. Random tests can run for long periods of time and generate large amounts of response data. The testbench needs to be able to check the responses as they occur so that the test can be stopped when an error occurs. In addition, the stimulus generator must provide an efficient interface for the random tests. It must define each of the parameters that the test writer might randomize and provide default constraints for each parameter. In the most simple form, a test might only need to specify how long to run, and the stimulus generator generates data based on the default constraints. The test writer should be able to override these constraints to create specific tests. The complexity in a constrained random test is not in the test but in the testbench.

Chapter 15

Advanced Testbenches
Using assertions and coverage

As the size and complexity of designs has increased, testbenches have become more focused on testing the design from its periphery. Testbenches have evolved to treating the design as a black box and testing it by applying stimulus at its inputs and observing its outputs with little regard to the specific implementation. This approach allows verification teams to verify large designs because it requires less intimate knowledge of the design. A black-box approach also facilitates reuse since less design-specific information is contained in the testbench. This focus on simplicity and reuse has resulted in testbenches that provide stimulus and checking capabilities but limited visibility of the implementation. Using a more visible or white-box approach to verification, however, eases debugging and provides knowledge on what has been tested. Assertions and coverage techniques, which have been touched upon throughout this book, provide a more white-box approach to verification. Today, black-box testbenches are combined with these white-box techniques to provide the efficiency, reuse, and flexibility that advanced verification requires.

ASSERTIONS

The use of assertions has become a hot topic in functional verification. What is an assertion and how does it apply to functional verification? Theoretical sources describe assertions as capturing the designer's intent and specifying intended behaviors. Whereas the practical view is that assertions are simply monitors placed in the design that identify actions within the design. These monitors can identify illegal behaviors and act as a supplement to the testbench checkers, or they can identify legal behaviors to help guide the verification process. Assertions are also often tied to formal or static verification techniques, which are discussed in another chapter. This chapter focuses on the practical application of assertions in a dynamic or simulation-based verification process.

Assertions address three basic verification issues. First, they address the issue of bugs being missed by the testbench and slipping through into the late stages of the verification process or even into the manufactured device. Catching a functional bug is a combination of stimulating the design to cause the bug to occur and checking the response to identify the consequences of the bug. The testbench is responsible for generating the stimulus to cause the bug to occur in a dynamic verification environment. Assertions increase the amount of checks within a design, making it more likely that a bug is found if it has been stimulated.

The second issue addressed is the time it takes to debug a failure in simulation. The process of debugging simulation failures is often described as peeling onions. The failure usually is first identified at the external interface of the design. The debugger must step back through each layer of the design to try to locate the cause. This can be very time-consuming and complex, depending on the size of the design and the location of the bug. Assertions provide checkers inside the design closer to the source of the bug, making debugging easier and faster.

Assertions also address the inefficiency of directing the verification process. An efficient dynamic verification process runs the test that has the highest likelihood of finding the next bug or verifying the most important feature in the design. This is difficult to do unless you know what has been tested in the past and what remains to be tested. Assertions can make the dynamic verification process more efficient by providing accurate and timely information about what has been tested.

Perhaps the easiest way to understand how assertions address many of today's verification issues is to look at the process of writing, running, and debugging a test and how assertions are used during this process.

Using Assertions in the Test Process

The first step in the test process is determining how to stimulate a desired function or feature of the design. You can write a test that directly proves the function to be correct or that simply stimulates the function with a wide range of stimulus to verify its operation. The effects of a bug can take many forms and, as the bug propagates through the design, the form can change. Test writers need to be aware of which types of checks are being performed by the testbench checkers. If the testbench checkers are not sufficient to catch a possible bug, the test writer might need to add checks. Even with random tests, you need to be aware of the propagation of bugs and the types of checkers available. Assertions address both these issues. Assertions detect bugs closer to the source, so less propagation is required. Assertions also provide a larger number of design-specific checks throughout the design, increasing the likelihood that a checker is in the correct place to catch a bug.

Once the test has been written, the next step is running the test. The biggest concern is knowing that the test is doing what was intended. In many cases, this is self-evident. In other cases, a bug in the design or a misinterpreted specification can cause the design to operate in an unintended manner, allowing the test to pass without verifying the intended function. One way to verify the operation is to review the waveforms of the simulation to make sure that the correct functionality was stimulated, but this can be a very time-consuming task. Another way is to place monitors in the design to identify when certain functions are being stimulated. Assertions provide monitors throughout the design, which can identify illegal as well as legal behaviors. You can also use assertions to identify that the test is not progressing and that the stimulus is not reaching the intended functionality.

After the test has been run, any identified failures need to be debugged. Debugging complex designs is often considered an art, but it is usually much more of a methodical approach. The first step in the debugging process is to try to understand as much about the failure as possible from the information provided by the test. The debugger traces back in the design to identify what might have caused the failure. This is done in the developer's mind, working with the knowledge he or she has of the design and test. In some cases, this information leads to identifying the bug quickly; in other cases, it simply narrows the possibilities down and gives the developer a starting point for finding the bug.

Experienced engineers know that spending time trying to find a bug by taking guesses at possible causes based on incomplete information is a waste of time. Instead, they use a more methodical approach of obtaining the necessary information, often described as "onion peeling." They first look at the most outer layer of information and then, based on that information, move down to the next lower layer and based on that information obtained, move to lower and lower layers. There are two approaches to this process. The first is starting from the beginning and tracing the stimulus into the design, verifying at each layer that the correct operations have occurred until an incorrect result is found. The other approach is to start at the end and trace the cause of the error back until the incorrect functionality is found. Once the area has been identified, it can be isolated and monitored to identify the exact cause of the bug. Then, a possible fix can be made to the design and the test reverified to see whether the bug has been fixed. It is often common for one bug to hide other bugs in the same functionality, so it is important to thoroughly verify the fix.

Assertions placed throughout the design can identify the behavior of the bug before it propagates and changes form. Assertions speed the process of localizing where a bug might occur by providing more information to the debugger earlier in the process. Instead of the developer rerunning the test and probing points in the design to localize the bug, assertions provide that information. Placing assertions around an area where a bug was found helps to reverify the bug fix and find any other bugs that might be hidden in the area.

The process of creating a test, running the test, and debugging failures is a repetitive task. Once a test has been debugged, the test developer returns to the next test and repeats the process. Selecting which test to create and run next is very important for determining the most efficient verification process. Randomly picking the next test or going in some arbitrary manner leads to inefficient verification that takes longer to complete and results in important bugs being found late in the development process. Advanced verification teams prioritize the order of tests to be run so that the debug process is efficient and the most important functions get tested early.

Verifying the functions in the order that they are used in the system, from input to output, allows the developer to more easily isolate a bug when a failure occurs. It is also important to prioritize the testing of the most important and most used functions early in the process. The goal is to identify bugs in important functionality first so that they can be fixed early in the development process. Often designs are required to tape out before the verification is complete. Prioritizing the most important functionality allows you to have the most confidence that the design will work if it has to tape out early. Prioritizing tests for debug and prioritizing by functionality might seem to conflict with each other. Advanced verification teams try to mesh these two goals. First, they prioritize for debug for early bring-up of the design to a level of sanity, and then they prioritize by functionality.

Assertions can help identify which areas of the design have been tested and which areas have been missed. Assertions provide coverage information at the implementation level, which helps identify the degree that structures and functions have been tested. We will talk more about using assertions as coverage monitors in the coverage section of this chapter.

Using Assertions

Assertions have been used in the past mostly by large development teams. These projects were usually very well staffed and had long development cycles. The basic use model was for assertions to be placed throughout the design either by the designer as the code was being written or by another engineer after the code was written. Assertions were placed using a specification language that best encapsulated the intended behavior of the implementation. The full suite of assertions would be simulated with the design each time a test was run. The benefit of this approach was that the code would be fully instrumented with a wide net of checkers to catch as many bugs as possible.

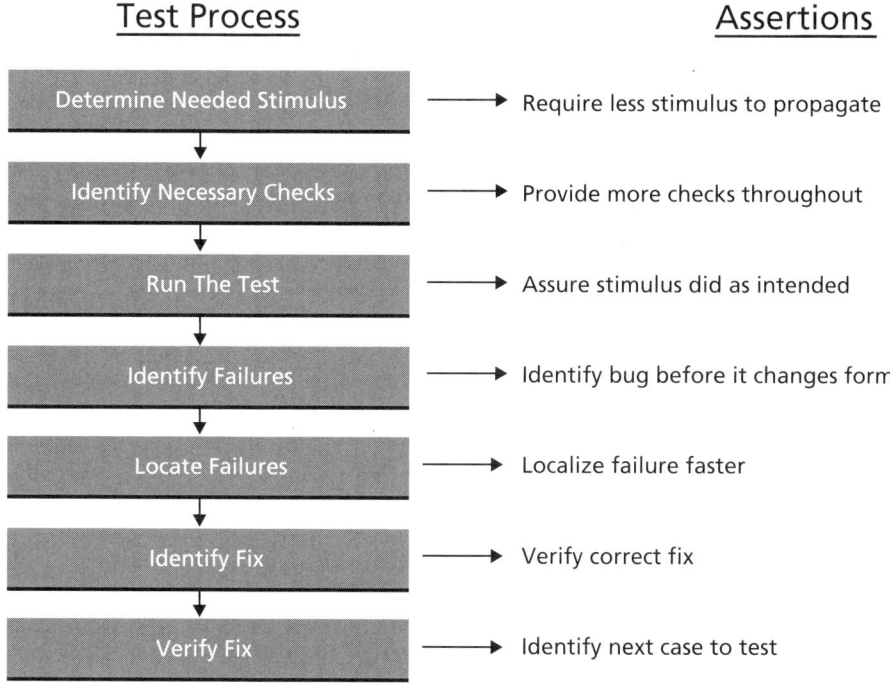

Figure 56. Assertions Facilitate the Test Process

Most development teams today lack the time and resources to create such an extensive net of assertions, nor do designers have the interest. Most designers do not see the value in using assertions because they believe that their code is correct and that assertions are a verification team's task. Recoding a design in a different language to capture its intent seems like a redundant task. The verification team usually lacks the implementation-specific knowledge and the time to put assertions in for the designer. So, if the designer does not create the assertions throughout the code, the older use model breaks down.

Advanced verification teams have learned to adapt their use of assertions to the realities of today's development process. These teams use assertions to address the specific needs of the process at each stage.

Assertions and the FVP

Architectural assertions can be used during system-level design. At this stage, architects or system-level verification engineers add assertions to verify features and architectural aspects of the design. These checks are often included in the architectural checks or transaction-level interface monitors described in the FVP. The assertions are usually written in the same language as the model and testbench so that speed is not impacted.

Assertions at the Block Level

At the block level, structural assertions can be used inside the design, and interface assertions can be used at the boundaries of the design. Using structural assertions at the block level is highly dependent on the designer. If the designer chooses to insert assertions, the most efficient path is for the designer to place assertions in the form of library elements instead of using a language to define the check. Using assertion pragmas or comments that allow the user to place the library in a shorthand manner facilitates this approach. Automated tools interpret these comments or pragmas and synthesize the assertion in the form of a library or language.

Structural assertions should be placed around design hot spots and in common bug-trap locations. Design hot spots are places in the design that are highly suspect of being incorrect based on past experience. Areas such as arbiters, state machines, or clock domain crossings are common areas for bugs. Designers should be encouraged to place assertions in the areas they believe to be suspect or where they have had problems in the past. Bug traps are places where the manifestation of bugs is most commonly seen. Placing assertions at places like a MUX, FIFO, or a handshake are often easy ways to catch the results of a bug in complex logic. A common rule of thumb many teams use is that if an assertion for complex logic cannot be written with a library or a few lines of language, you should use bug traps to catch the effects instead.

Once the code for a block is complete, the designer or a verification engineer can add interface assertions to the block. Assertions for standard interfaces can easily be added using a library approach. If the interface is not standard and requires some unique checking, the assertions should be written in a standard language. Interface assertions should be added at the interface between major functional blocks and between blocks created by different designers.

Block-level simulations should be run with both structural and interface assertions. Assertions might not be of value in the very early stages of bring-up where clocks may be incorrect or the design not reset correctly. Most teams choose to disable assertions until the design has reached a stable reset or initialized state. The first stages of debugging can cause numerous structural assertions to fire due to the same simple bug. Teams might choose to ignore structural assertions until the first few tests have passed.

Assertions and Chip-Level Verification

After individual blocks have been verified in isolation, they are integrated together and verified as a full chip. Structural and interface assertions of the individual blocks are integrated along with the design. Additional interface assertions are placed at the primary inputs and outputs of the design and at any missing internal interfaces. At this time, architectural assertions developed at the FVP are also integrated if they are not already part of the chip test environment. If an FVP was not used or if additional architectural assertions are required, they can be added using a standard language.

Simulation with assertions at the chip level is similar to the block level. Assertions should be turned off during initialization, and structural assertions might be ignored during basic bring-up. Fewer assertions fire during chip-level simulation since they have already been thoroughly exercised during block-level simulation, so each violation should be examined carefully. If there are a large number of assertions in the design, performance could be impacted. The team can turn off structural assertions if performance is unacceptable. The team might also choose to use a hardware accelerator, in which case all the assertions might need to be synthesized so that they can be accelerated with the design.

Assertions and System Verification

Many teams stop using assertions once they move from a testbench-based verification environment to a real-world system verification environment. Teams often remove assertions when using an emulation system or an FPGA because of speed and capacity requirements. While it is important to focus on speed and capacity during system verification, assertions can provide important visibility that is often lost in emulation of FPGA-based systems. These systems provide little or no internal visibility to help debug failures. Assertions should be included with the design in the emulation or FPGA system to provide this needed visibility.

Flexibility and Reuse

Advanced verification teams do not get hung up on the mechanism for placing assertions in the design. The teams use whatever is the best mechanism for that point in the process. The result may be a design that has assertions in various different formats, libraries, or languages. Teams should not get tied down to one format. The verification platform should be able to support these various forms in a unified manner.

It is also important to try to reuse assertions throughout the process. Assertions created at the block level should also be used at the chip level and the system level. The team should not get tied into one proprietary way of specifying assertions, because this leaves assertions fragmented from stage to stage. Assertions should also be used in software simulation as well as hardware simulators and formal verification tools. Care should be taken to ensure that the time and effort expended in creating assertions is amortized across all these areas.

COVERAGE

One of the most common issues verification teams face is determining when they have done enough verification to be confident that the design is ready for production. Verification teams have attempted to use coverage as a metric for determining completeness, but often come to the realization that these techniques fall short of the original goal. Advanced verification teams have learned that the coverage techniques and tools used today simply provide raw information that the user must correlate and comprehend before any actions can be taken or conclusions made. The information is also incomplete or inconclusive. Still, verification teams have found that the information that coverage techniques provide is valuable in guiding the verification process in the most efficient manner. In this section, we will explore coverage techniques used by advanced verification teams.

Coverage cannot provide an answer to the question of completeness, but it can provide some of the data to help make that determination. Advanced verification teams understand that the verification process is often very similar to risk management. Before making an investment or placing a bet, one should assess the risk of losing money and compare that to the possibility of being rewarded. In a similar manner, when a development team decides they are ready to tape out their design, they should assess the risk that there is a bug in the design that will make the device unusable. This risk must be weighed against the possible rewards of getting to market sooner. Coverage tools provide the type of information that, along with experience and proper processes, enables the team to make a fair and accurate risk assessment.

Coverage can help keep the verification process on the most effective and efficient path. The verification process of a complex device can last many months or years. Verification plans and strategies are usually developed early in the process and might not be modified unless there is a major change in the project's direction. Advanced verification teams have found that it is good practice to periodically check where they are in the process and reevaluate if the original strategies are still the correct ones. High-level inexact data provided by code coverage or stimulus coverage tools can be enough to identify which areas of the design are receiving the most verification and which they might want to concentrate on more.

Perhaps the most important verification issue addressed by coverage is the one that is most often overlooked. Coverage can be used to identify areas within the design that have not been stimulated and, therefore, may be hiding potential bugs. The previous section on assertions discussed the process for finding bugs within a design. The first step in that process was generating stimulus that stimulates the design in a way that manifests the bug. Verification teams often find that the reason a bug has slipped through the verification process is that they never verified that operation. Coverage information can be used to find bugs that are missed in the verification process.

Using Coverage

Using coverage within a verification environment consists of three stages: identification of goals, simulation, and analysis. The first stage in any measurement process is determining what it is you are attempting to measure. Advanced verification teams set coverage goals as part of their verification strategy. The verification team develops a test plan that identifies the functionality to be tested and a strategy for testing each function thoroughly. Coverage goals are set for each function. These goals list which metric is important to obtain to verify that the function has been tested. The metric could be as simple as a certain signal being asserted a number of times or as complex as a series of protocol sequences.

Once the test plan is completed, a coverage model that details the functional coverage goals, along with stimulus coverage goals and goals based on the team's experience, is developed. The coverage model identifies how each of these goals is measured. The team uses the test plan and coverage model to develop the testbench for stimulating and testing the design. The team might track stimulus generation, increase interface monitors, or add internal monitors to identify certain coverage goals. Some of the coverage goals are attained through using automated coverage tools, such as code coverage or structural coverage tools.

The coverage instrumentation is included as the entire test suite is simulated. The monitors collect information from every test and correlate them into a single database for analysis. The entire test suite should be simulated before analysis is started. Time can be wasted identifying holes in the verification process that are already covered in tests that have not been run. It is also important to only run tests that pass. Tests that report incorrect functionality should not be included in the design, since the information might lead to incorrect analysis.

One of the most difficult decisions is deciding when to measure coverage. If you start too early, you waste time chasing holes that will be covered in future tests. If you start too late, the data is usually too late to do any good. Ideally, you would run coverage once all the tests are complete, all the tests are passing, and the design is free of bugs. In reality, this condition rarely occurs. The verification team must pick a point in the process where they feel the design and test environment are fairly stable. At this point, a snapshot should be taken of the entire environment, and a separate coverage simulation run should be done. Having a separate stable coverage run makes analysis easier and limits the performance impact of coverage collection.

The final phase is to analyze the coverage information collected. The first information to analyze is the specific coverage goals for each function listed in the test plan. This identifies any tests that are working incorrectly or that have been missed from the test plan. If any coverage goals are not met, the team should create new tests and resimulate the design until all the goals are met. The next step is the cross- correlation of the functional coverage goals. The coverage goals met during the first step are crossed with each other and with related functional verification information, such as stimulus coverage. This step checks that not only were the intended functions verified but they were verified in conjunction with other related activities within the design. If holes are discovered in the cross-correlation, the team should create new tests to cover these cases and resimulate the design until the goals are met.

The first two phases of the coverage analysis process verify that the functionality that the team knows about has been tested properly. The last phase is to identify coverage holes in areas that the team may not be aware of. This phase uses automated coverage to identify missed areas or areas added during the development process without updating the test plan. This information is often inaccurate or incomplete, so the verification team should not begin this analysis until they are sure that everything that they know needs to be verified has been verified. If holes are identified in this stage, the verification team should create new tests to cover these areas and resimulate the design until the goals are met.

The coverage process is an iterative process of identifying missed goals or stimulus holes and addressing them in a prioritized fashion. Metrics can be collected along the way to mark the progress and to give management an indication of confidence in the design. In most cases, the collection and analysis of coverage information is not started until the test suite is near completion and the design is stable. Measuring coverage too early in the process can result in incomplete information leading to incorrect assumptions.

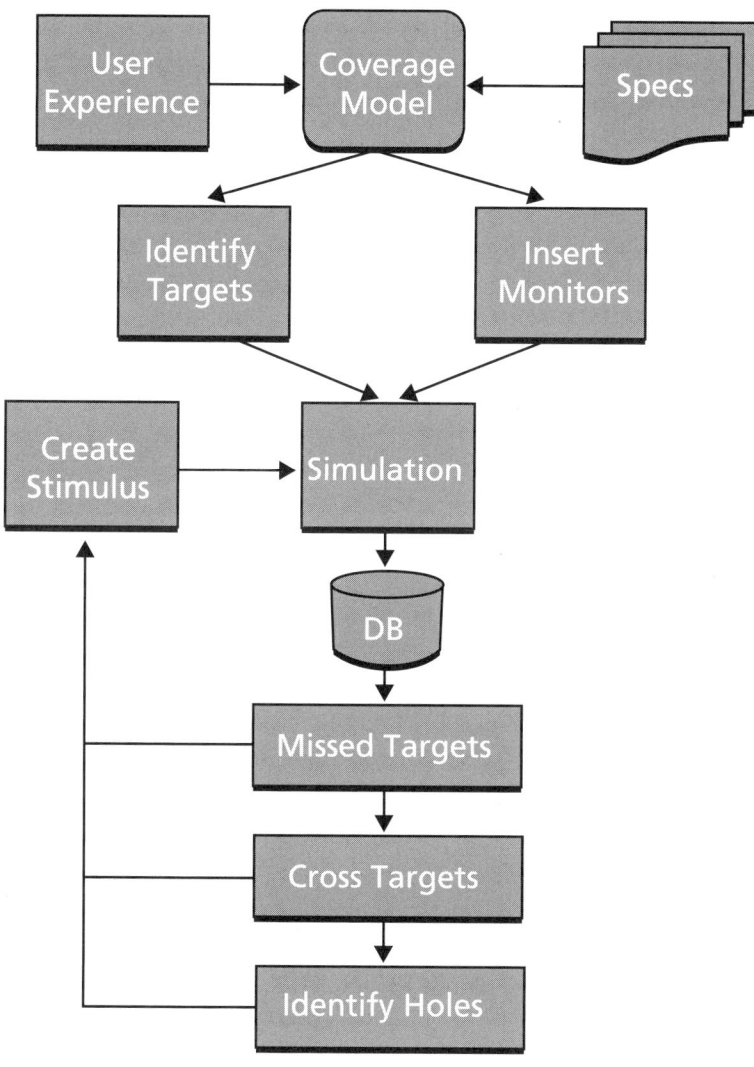

Figure 57. The Coverage Process

Filling Coverage Holes

An often overlooked aspect of coverage is how you generate new stimulus when you identify a coverage hole. In some cases, the reason a goal was missed or an area was not covered is simply an oversight, such as forgetting to set a mode bit. These cases can be covered by correcting the oversight. More often, the reason is the difficulty in causing that goal or area to be stimulated. The difficulty in generating stimulus could be caused by logic being buried deep in the design, or that the logic requires a complex and rare sequence of events to stimulate or many cycles of setup to occur to bring the design into a state where the area can be stimulated. Each of these cases creates a challenge for the verification engineer.

A verification engineer has several options to choose from when attempting to generate stimulus to cover a difficult coverage hole, such as writing a directed test that targets the specific logic or functionality. However, writing a directed test to cause specific internal interactions to occur can be difficult and requires in-depth knowledge of the implementation. If the verification environment includes random tests, the engineer could rerun these tests for longer periods of time or with different seeds, hoping that eventually the test covers the intended area. Depending on how large the design is and how constrained the random stimulus generation is, the chances of this working can be quite low. Constraining the random stimulus generator to narrow the focus to the area of the design that needs coverage could improve the odds that the area gets covered, but might not help if the effects of variance in the stimulus are limited to the interfaces of the design.

Another option is to use automation to create stimulus to target a coverage hole. Some advanced verification teams use formal verification tools to generate test cases to stimulate the intended area. These teams place a property or assertion within the design associated with the coverage target. Instead of using the tool to prove that the case can never happen, they use it to generate cases that can make it happen. These test cases are then used to derive a directed test to cover the intended coverage goal. This technique requires knowledge of using a formal verification tool and requires the setup associated with using a formal tool, such as constraining the design and defining initial states.

Advanced verification teams have also used coverage-directed stimulus generators, which read in coverage information and attempt to generate stimulus to cover the areas not yet covered. These tools have been limited to narrow coverage information, such as stimulus coverage. Currently, the goal of automatically identifying and stimulating functions or structures within a design has only been used in research projects.

REACTIVE TESTBENCHES

One way to create stimulus to address coverage holes is to use run-time coverage information. The coverage process detailed earlier relies solely on using post-processed coverage information that is not analyzed until all the simulations complete. Run-time coverage provides dynamic information that can be used as the tests are running. It can be used for event notification and information collection. You can use run-time coverage information with each of the different stimulus generation techniques for addressing coverage holes. Directed tests can utilize the notification of internal events to direct stimulus to the targeted area. Random tests can use the collected information to change constraints on-the-fly. Coverage-directed stimulus generators can use the information to identify coverage holes and to direct stimulus to targeted areas.

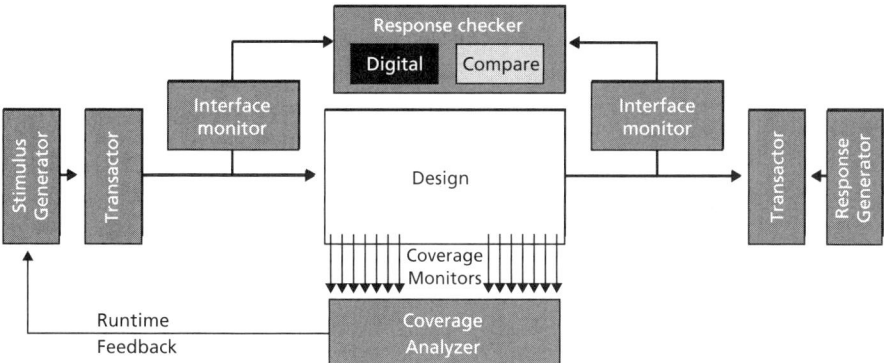

Figure 58.   Reactive Testbench

Run-time coverage information facilitates the development of reactive testbenches. Reactive testbenches enable test writers and stimulus generators to change the actions or characteristics of a test while it is running. Creating a reactive testbench requires the ability to observe the behaviors of the design and to react to those behaviors. Testbench environments have always had the ability to probe into a design to monitor internal signals. This has mostly been done to check the design's response. The limitation of probing internal signals is that the engineer must know of the signal's existence and know how to identify it. Changes in the design or in the hierarchy often lead to the signals being removed or misidentified. Coverage goals and functions are not as easily correlated with internal signals as response checks are. Thus, it has been more difficult to use simple signal probing to monitor designs for reactive testbenches.

Advanced verification teams have used assertions to overcome this limitation. Structural and functional assertions placed inside a design translate internal signal information into functional events that can be monitored and reacted to. Assertion tools handle the translation of signal names and hierarchies as the design and testbench mature. So, instead of the test writer having to identify the read, write, and location counters of a FIFO to determine when the FIFO is nearly full, a simple FIFO assertion identifies them. Using assertions as coverage monitors is often overlooked. One approach to creating a reactive testbench environment is developing and using a library of structural assertions that act as checks and monitors. These assertions can be placed on elements, such as FIFOs, FSMs, arbiters, and MUXes, to check for illegal operations and to identify that the elements are operating functionally. These assertions provide the test writer with an interface to the internals of the design that is not directly tied to the implementation.

Once a reactive testbench has the ability to easily observe the design, reacting to those observations can be developed. This reaction can be as simple as stopping the test. Often it is difficult for test writers to know when the test has caused the intended behavior within a design, so they run the test for a maximum number of cycles to be assured it has happened. This can be a waste of simulation time. Instead, in a reactive testbench, the test can run until it has been notified that the intended behavior has occurred and the effects have been verified. The reactive testbench may also be able to track the propagation of stimulus through a test so that the test developer can be sure the effects have propagated to a checker. Reactive testbenches can also detect when intended behavior is not occurring so that the stimulus can be changed or the test terminated.

Reactive testbench environments can also make the development of complex directed tests much easier. Complex directed tests often require the test writer to set up complex interactions within a design, such as filling a FIFO on the same cycle as a state machine changes state. Creating complex interactions requires the test writer to understand the details of the implementation to know what exact stimulus to provide to the design. A reactive testbench can provide information to the test writer to make this process easier. A FIFO monitor can indicate when the FIFO is nearly full, and a FSM monitor can identify the current state of the FSM. Using this information, the test writer can send data until the FIFO is nearly full, then wait until the FSM is in the desired state before sending the next stimulus to fill the FIFO.

In addition to simply reacting to the behaviors of the design, you can use reactive testbenches to monitor what the design has already done and attempt to do something new. Automating the coverage process has long been a goal of advanced verification teams. While it may be considered a science project today, it is worth investigating the problem to better understand the capabilities of coverage tools and reactive testbenches. Today, the coverage process requires manual intervention for selecting goals, identifying holes, and creating stimulus. Selecting goals will probably always require some manual intervention, but automation can help identify possible goals and provide simple coverage goals, such as code coverage or stimulus coverage.

A completely automated coverage process requires a reactive testbench to perform the identification of coverage holes and the creation of targeted stimulus while the test is running. Identifying functional coverage targets that have not been stimulated is straightforward. The assumption is that every goal specified by the user is of high value, so it must be covered. Identifying general coverage goals is more complex. Today, an engineer reviews coverage data provided by a code or functional verification tool to identify which are real holes and which are not. General coverage information might identify areas of the design that are uncovered that are not intended to be covered, such as test logic. A verification engineer knows that the reason a certain area of the design is not covered is because those tests have not been run yet. Automated tools cannot infer this information, so automatic identification might be limited to user-specified targets.

The automatic creation of stimulus to target coverage holes in a design will probably require the combination of several different verification technologies. Today, some stimulus generation tools can monitor the stimulus they have already sent and adjust the generation to cover stimulus that has not yet been sent. This approach to automated generation is the first step in a more complex problem. As noted earlier in the chapter, stimulus coverage does not provide enough detailed information about how the design was stimulated to be very useful for most designs. Most coverage holes are not at the interfaces of a design but are buried deep in the functionality. The process of targeting these internal functions is to step back from the functionality layer by layer to determine which sequence of operations needs to occur for the target to be stimulated. The ability to do this traversing of the design is limited to formal verification and symbolic simulation techniques. So, the solution to this problem will probably be a combination of verification techniques from simulation, constrained randomization, formal model checking, and symbolic simulation.

Advanced verification teams have begun to use run-time coverage information and reactive testbenches. One issue these teams have run into is the difficulty in keeping the tools up-to-date with coverage information. Most complex design projects have hundreds or even thousands of different tests to run. If an environment or test is going to use coverage information from these past tests, there is often a chicken-and-egg type problem. The user has to run all the other tests to collect the data so that the next test can use it. If a test fails or a change is made to the design or testbench, the tests have to be repeated before the information can be used again. Most teams run their tests simultaneously on server farms, which requires close coordination and synchronization of different active processes. The lesson learned from these advanced teams is to fully understand the intended use model before embarking on using an advanced run-time coverage environment.

Chapter 16

Hardware-Based Verification
Advantages of hardware-software co-verification

As more and more electronic products have software content, designers are faced with serious project delays if they wait for first silicon to begin software debugging. It also means that a serious system problem might not be found until after first silicon, requiring a costly respin and delaying the project for two to three months. Increasingly, designers are turning to hardware-software co-verification—concurrently verifying hardware and software components of the system design—to meet demanding time-to-market requirements. At a minimum, this means starting software debugging as soon as the IC is taped out rather than waiting for good silicon. But even greater concurrency is possible. In many cases, software debugging can begin as soon as the hardware design achieves some level of correct functionality. Starting software debugging early can save two to six months of product development time. There are a variety of approaches to hardware-software co-verification. This chapter addresses accelerated co-verification, since the complexity of software in today's electronic products precludes adequate testing with the performance of a software simulator.[1]

ACCELERATED CO-VERIFICATION

There are additional benefits to starting software verification prior to freezing the hardware design. If problems are found in the interface between hardware and software components, designers can make intelligent trade-offs in deciding whether to change the hardware or software, possibly avoiding degradation in product functionality, reduced performance, or an increase in product cost.

[1.] This chapter is based on Cadence Design Systems' white paper "Accelerated Hardware/Software Co-Verification" by Ray Turner, March 2004.

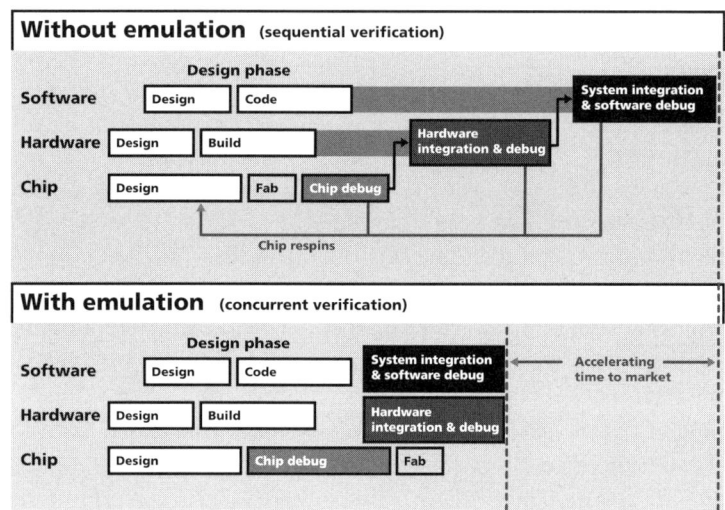

Figure 59.   Emulation Allows Concurrent Chip, Board, and Software Verification

Usually, a custom IC is to be used with a standard microprocessor, which is running the software being developed. The IC, the software driving it, and the board and system have to be verified. Typically, the IC is verified using simulation, accelerated simulation, and, especially for larger ICs, in-circuit emulation, using test vectors, testbench programs, and live in-circuit data. The board and system, including software, are usually verified in a live use context, frequently augmented with special test equipment. Less frequently, microprocessor suppliers are developing a new processor or variant either as a standalone device or as a core. In this case, the processor must also be verified for software compatibility and system-level interface.

An important factor in selecting a co-verification approach is the type of model available for the processor. It can be a physical component, such as a standard microprocessor put on a board or a bond-out core for a processor core included in an IC. It can also be an RTL model of the processor. In general, in-circuit emulation provides the highest performance possible—several orders of magnitude higher than any simulation approach. Emulation also allows verification of the design in a real-world environment with live data. Testing a design in the context of actual data, with thousands of times the volume of test data, provides exceptionally high confidence in design correctness. If only an ISS model is available, acceleration is possible, but overall performance is reduced by the speed of the ISS model, perhaps by as much as one or two orders of magnitude.

We will now look at three approaches to accelerated hardware-software co-verification. They all support designs with multiple processors. These approaches are listed in order of increasing performance and use the following:

- An ISS with a software simulator and an accelerator or emulator
- An RTL processor model and an accelerator or emulator
- A physical model of the processor (bond-out core) and an emulator

Using an ISS, Software Simulator, and Accelerator/Emulator

In this approach, the logic simulator models the IC and other hardware components, except for the processor and memory. Processor simulation is done by an ISS, and memory is modeled by workstation memory. An industry-standard programmable language interface (PLI) connects the simulator to the ISS. In operation, the system software is executed by the ISS on the workstation. The ISS typically can execute several thousand instructions per second. When an I/O instruction or memory-mapped I/O access to the IC is performed, the ISS passes the I/O to the simulator, which handles any non-synthesizable code in the design and interfaces to the accelerator, which is accelerating all the synthesizable code in the design. Any resulting changes in IC outputs are passed back to the ISS. Overall performance depends greatly on the amount of I/O being done by the software due to the overhead of communicating between the ISS and the accelerator. An important benefit of this approach is that industry-standard software debugging tools are used, which can be more productive for investigating software problems. An ISS model of the processor is needed, but these are generally available and typically supplied by the processor vendor as part of the software development toolset.

The steps in using co-verification with an ISS, software simulator, and accelerator are:

1. Compile the software into a ROM code file.
2. Compile the hardware design for the accelerator and download.
3. Start the simulator and the ISS.
4. Debug with software and accelerator debug tools.

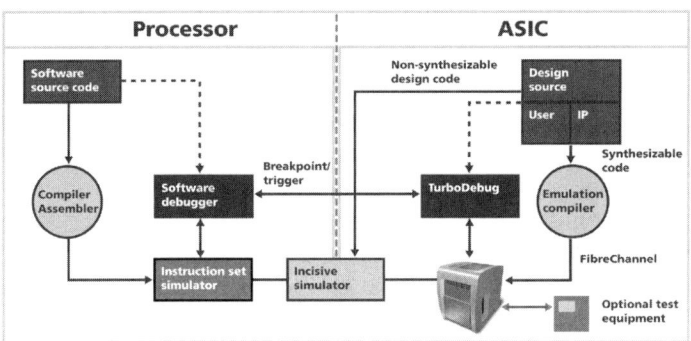

Figure 60. Using an ISS, Software Simulator, and Accelerator

Using an RTL Processor Model and Emulator

In this approach, an RTL model of the processor is substituted for the ISS model. The RTL model is mapped into the emulator along with the IC design. The entire system is modeled in the emulator and runs at full emulation speed—usually ten to a hundred times faster than the ISS approach. You can connect a software debug monitor to provide the familiar software debugging environment. Thus, the software and hardware engineers each use the debug environment they are most familiar with, thereby increasing debug productivity.

Since this approach can be in-circuit and in-system, testing can take place with live data in as real world an environment as possible. This approach is the only way to gain the high confidence that comes with testing a design in a real environment with real data. It is hard to overestimate the value of in-system testing. Over and over again engineers talk about finding bugs in this way that they could not possibly have foreseen or tested for in a simulation environment. The only substantial difference between testing with emulation and testing with first silicon is that in emulation the target environment must be slowed down to emulation speeds and, therefore, provides lower performance than actual silicon, but with the advantage of complete visibility into the design and a comprehensive debugging environment that first silicon does not offer.

The steps in using co-verification with an RTL model and incisive emulator are:

1. Compile the software into a ROM code file.
2. Compile the hardware design for the emulator and download.
3. Plug the emulator and software debugger into the target system, if used.
4. Debug with software and emulator debug tools.

Using a Physical Model of the Processor and an Emulator

In this approach, the RTL model is replaced by a physical model: a bond-out core. Performance is similar to the RTL approach, but less capacity is needed in the emulator. Aside from the obvious performance advantage, this approach also allows microprocessor in-circuit emulators (MP-ICE) to be used for software development and debugging. These provide for rapid download of the code into the target system and sometimes provide additional functionality, such as hardware breakpoint detection, watched variables, and/or a logic analyzer. They might also provide more functionality than simpler debug monitors, for example, disassembly of the executed code to assist debugging and source code execution control, such as single-stepping and breakpoints. Hardware and software debugging tools can be easily cross-coupled for coordinated debugging, when needed.

The steps in using co-verification with a physical model, emulator, and MP-ICE are:

1. Compile the software into a ROM code file.
2. Compile the hardware design for the emulator and download.
3. Plug the emulator and MP-ICE into the target system (or IP chassis).
4. Cross-connect the emulator and MP-ICE for coordinated debugging.
5. Download the ROM code file with the MP-ICE.
6. Debug with software and emulator debug tools.

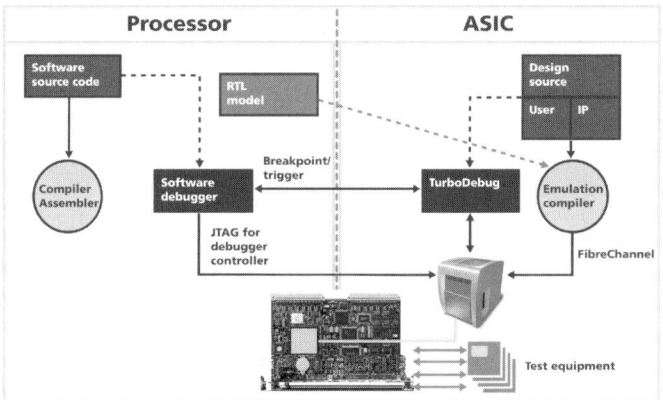

Figure 61.   Using a Physical Model, Emulator, and MP-ICE

In some target-based approaches, a Real-Time Operating System (RTOS) might be running in the processor or a debug monitor might be running (sometimes called a Resident System Monitor (RSM)). These provide a communications path (RS-232 or Ethernet) back to a workstation running the software debugger. In both the MP-ICE and RTOS cases, the software and hardware debugging environments can be synchronized so that hardware-software interface issues can be debugged conveniently. The breakpoint or trigger systems of the emulator and MP-ICE are cross-connected such that the emulator's logic analyzer trigger is one of the MP-ICE breakpoint conditions, and the MP-ICE breakpoint trap signal is set as a emulator logic analyzer trigger condition. If a software breakpoint is reached, the emulator captures the condition of the IC at the same moment. If an IC event occurs that triggers the logic analyzer, the software is stopped at that moment. This allows inspection of the hardware events that led to a software breakpoint or of the IC operation resulting from executing a set of software instructions. This kind of coordinated debugging is extremely valuable for understanding subtle problems that occur at the hardware-software interface.

Comparing Approaches

The table below summarizes the trade-offs of the three approaches explained above.

Table 4. Comparison of Approaches

Approach	Type of Model	Debug Environment	Performance	Level of Software That Can Be Verified
ISS, Simulator, and Accelerator	ISS	ISS debugger and logic simulator	Medium-High	Drivers and Diagnostics; Small OS
RTL model and Emulator	RTL	SW debugger and TurboDebug	Very High	UNIX, Windows, RTOS, and applications
Physical model and emulator	Physical	MP-ICE, RTOS, and TurboDebug	Very High	UNIX, Windows, RTPS, and applications

There are several factors to take into account when determining which approach is best for your project. One factor to consider is the performance required to meet your objectives. Another is whether you are going to begin software debug before or after tapeout. The amount of software that you want to verify is another consideration. If you only want to verify very little software before working silicon is available, use logic simulation and an ISS. For a moderate amount of software, use an ISS and an Incisive simulator and accelerator. If you want to verify a lot of software, use RTL or a physical processor model and emulation.

INCORPORATING CO-VERIFICATION INTO YOUR DESIGN ENVIRONMENT

One of the most significant factors in implementing hardware-software co-verification is the corporate culture and organization regarding hardware and software developers. The ideal is a project team in which hardware and software engineers report to a single project lead or manager and work together in a fully collaborative way to create an optimal hardware-software system. An accelerator or emulator can be shared very effectively in a multi-user environment. The capacity of a single emulator can be shared among eight users for BIOS and driver software development. Alternately, the entire system capacity can be used when verifying the complete design. With multiple systems, you can support many simultaneous software developers with a very high performance verification and debugging environment.

Looking at the cost of a verification solution versus the cost of making a mistake can be instructive. The costs of making a mistake include the cost to do a respin of the IC and the cost of being three months late to market (the average time that can be saved if you start software debugging before tape-out). For rapidly changing consumer markets, the lost opportunity cost can easily be tens of millions of dollars. There are additional benefits from using co-verification. For example, it is very helpful if the diagnostics are running when the IC comes back from fabrication. They can be used to do focused testing of specific parts of the design. Without working diagnostics, you end up doing ad hoc testing of the whole IC at once—a hit-and-miss proposition.

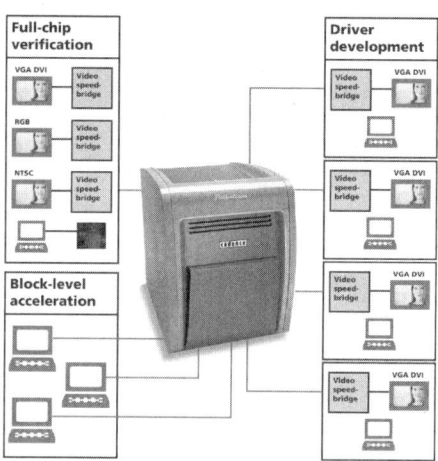

Figure 62.   Emulators Can Be Shared in a Multi-user Environment

With today's complex ICs, acceleration and emulation are practical necessities to verify designs and software in a complete system environment with real data. Software content of electronic products is increasing exponentially and is most often the pacing item for product completion. Using acceleration or emulation for hardware-software co-verification takes advantage of the investment made in the emulator and shortens product cycles by several months. Emulation as a vehicle for hardware-software co-verification provides by far the highest performance available for this critical task, along with real-world data for comprehensive system testing.

Appendix 1

Resources

Bergeron, Janick. *Writing Testbenches—Functional Verification of HDL Models*, 2nd ed. Boston: Kluwer Academic Publishers, 2003. ISBN: 1402074018

Foster, Harry, Adam Krolnik, and David Lacey. *Assertion-Based Design*, 2nd ed. Boston: Kluwer Academic Publishers, 2003. ISBN: 1402074980

Grotker, Thorsten, ed. Stan Liao, Grant Martin, Stuart Swan. *System Design with SystemC*. Boston: Kluwer Academic Publishers, 2002. ISBN: 1402070721

Haque, Faisal, Khizar Khan, and Jonathan Michelson. *The Art of Verification with Vera*. Verification Central, 2001. ISBN: 0-9711994-0-X

Meyer, Andreas. *Principles of Functional Verification*. Newnes, 2003. ISBN: 0750676175

Muller, Wolfgang, Wolfgang Rosenstiel, and Jurgen Ruf, eds. *SystemC: Methodologies and Applications*. Boston: Kluwer Academic Publishers, 2003. ISBN: 1402074794

Palnitkar, Samir. *Design Verification with e*, 2nd ed. Boston: Kluwer Academic Publishers, 2003. ISBN: 0131413090

Sutherland, Stuart. *Verilogs 2001: A Guide to the New Features of the VERILOG Hardware Description Language*, 1st ed. Boston: Kluwer Academic Publishers, 2002. ISBN: 0792375688

Glossary

Acceleration-on-Demand The ability to move from a software simulation-based test environment to a hardware-accelerated, simulation-based test environment.

Algorithmic-based Digital Design A digital logic design that is directly developed from a algorithm or protocol and does not contain control-based operations.

Analog Behavioral Model A model of an analog circuit that represents the behavior of the implementation, but does not include the implementation-specific information.

Application Assertion An assertion used to specify an application-specific architectural property, such as fairness of an arbiter.

Application Coverage A measurement of the percentage of application coverage monitors that have measured an event.

Application Coverage Monitor A device to monitor the number of times an application-specific event has occurred.

Architectural Checks Checkers that verify the correct functional and performance operation of the FVP.

Assertion A codified representation of a designer's or architect's intent when creating a design. Assertions specify a property or behavior in a structured manner that can be verified to be correct.

Bottom-Up Development An approach to development starting at low-level implementation blocks and integrating the blocks together to form system-level representation.

Control-based Digital Design A digital logic design that is developed from a specification and not strictly based on a algorithm or protocol.

Design Hierarchy The naming of the design's hierarchical levels in a system. A system is made up of subsystems, which are made up of design blocks, which are made up of design units.

Device Under Verification (DUV) The block, system, or subsystem being verified.

Emulation A system verification technique where an implementation of a design is mapped into a hardware device that emulates the operation of the design at faster speeds than simulation. The device provides standard interfaces to connect the design to real-world interfaces.

Functional Virtual Prototype (FVP) A golden functional representation of the complete DUV and its testbench.

Implementation-Level Model A functional model of the design in which the structure and communications interfaces are defined to be implementation-specific. These models are often referred to as Register Transfer Level (RTL) models.

Interface Assertion An assertion used to specify the protocol and handshaking of an interface between two blocks.

Interface Coverage A measurement of the percentage of interface coverage monitors that have measured an event.

Interface Coverage Monitor A device to monitor the number of times an interface event has occurred.

Interface Monitors A testbench component that passively monitors an interface looking for signaling and protocol errors.

Mixed-Signal Design Designs that combine analog and digital logic.

Response Checker A testbench component that compares the output of the DUV to the expected response to verify correct operation.

Response Generator A testbench component that responds to requests made by the DUV.

Single-Kernel Architecture The ability to natively support all design and verification languages from the same simulation engine.

Stimulus Generator A testbench component that creates stimulus and sequences its delivery to the DUV.

Structural Assertion Used to specify the operation of low-level implementation-specific structures in a design.

Structural Coverage A measurement of the percentage of structural coverage monitors that have measured an event.

Structural Coverage Monitor A device to monitor the number of times an implementation structure event has occurred.

System-on-Chip (SoC) An integrated circuit with an on-board processor, memory, and one or more standard interface blocks or application-specific blocks.

Top-Down Development An approach to development starting at a high-level representation and partitioning down to lower level implementation blocks.

Transaction A unit of information abstracted from a lower signal-level representation that is used to represent an information transfer separate from the mechanism of transfer.

Transaction-Level Model (TLM) A functional model of the design in which communications interface is in the form of transactions.

Transaction Taxonomy A classification of the types of transactions used throughout a design.

Transactor A testbench component that converts different levels of interface abstraction, such as signal-level interface to transaction-level interface.

Index